Evolution, Creationism, and Intelligent Design

Recent Titles in
Historical Guides to Controversial Issues in America

Evolution, Creationism, and Intelligent Design

Allene Phy-Olsen

Historical Guides to Controversial Issues in America

 GREENWOOD

AN IMPRINT OF ABC-CLIO, LLC
Santa Barbara, California • Denver, Colorado • Oxford, England

Library of Congress Cataloging-in-Publication Data

Phy-Olsen, Allene.
    Evolution, creationism, and intelligent design / Allene Phy-Olsen.
        p. cm. — (Historical guides to controversial issues in America)
    Includes bibliographical references and index.
        ISBN 978-0-313-37841-6 (hardcopy : alk. paper) — ISBN 978-0-313-37842-3 (ebook)
    1. Evolution (Biology)   2. Creationism.   3. Intelligent design (Teleology)
    I. Title.
    QH367.3.P485   2010
    202'.40973—dc22   2010009743

ISBN: 978-0-313-37841-6
EISBN: 978-0-313-37842-3

14 13 12 11 10   1 2 3 4 5

This book is also available on the World Wide Web as an eBook.
Visit www.abc-clio.com for details.

Greenwood
An Imprint of ABC-CLIO, LLC

ABC-CLIO, LLC
130 Cremona Drive, P.O. Box 1911
Santa Barbara, California 93116-1911

This book is printed on acid-free paper ∞

Manufactured in the United States of America

For Frederick B. Olsen, with whom I have spent the happiest years of my life

Religion is the vision of something which stands beyond, behind, and within, the passing flux of immediate things, something which is real, and yet waiting to be realized, something which is a remote possibility, and yet the greatest of present facts, something that gives meaning to all that passes, and yet eludes apprehension, something whose possession is the final good, and yet is beyond all reach, something which is the ultimate ideal, and the hopeless quest.

The immediate reaction of human nature to the religious vision is worship. Religion has emerged into human experience mixed with the crudest fancies of barbaric imagination. Gradually, slowly, steadily the vision recurs in history under nobler form and with clearer expression. It is the one element in human experience which persistently shows an upward trend. It fades and then recurs. But when it renews its force, it recurs with an added richness and purity of content. The fact of the religious vision, and its history of persistent expansion, is our one ground for optimism. Apart from it, human life is a flash of occasional enjoyments lighting up a mass of pain and misery, a bagatelle of transient experience.

—Alfred North Whitehead, *Science and the Modern World*

There is grandeur in this view of life, with its several powers having been originally breathed by the Creator into a few forms or into one; and that, whilst this planet has gone cycling on according to the fixed law of gravity, from so simple a beginning endless forms most wonderful and most beautiful have been, and are being evolved.

—Charles Darwin, *On the Origin of Species*

Contents

Introduction

We have a problem! Many politicians, social scientists, educators—and, most of all, parents—agree that the U.S. public schools are in crisis. American students have lower scores in mathematics and the sciences than comparable students in other industrialized nations. Scientific associations and educators complain that parental attitudes and school board restrictions frequently hinder their efforts to teach current scientific developments and theories. Textbook publishers, with a keen eye on their markets, often ignore or treat controversial issues in the sciences ambiguously. On the other hand, many parents and concerned citizens believe the public schools are ignoring their values and slighting their religious beliefs. As a result, they feel that public schools, without standards or ideals, have become dangerous places where drugs and violence flourish.

Increased home schooling and private education, which can be prohibitively expensive for the less affluent, are one solid result. When the most responsible citizens withdraw their support for the public schools, or when there is no alternative path to education for most families, the acknowledged problems are only perpetuated. Conflicts of class and race may be intensified. The public schools, which once successfully instilled patriotism and promoted "truth, justice, and the American way" in children from many different ethnic backgrounds, have fallen into decline and are no longer the institutions that were once the pride of their communities.

In past centuries, theology and science were usually regarded as two complementary paths toward the same goal: the pursuit of knowledge, truth, and an understanding of the ways of the Creator. Muslim astronomers charted the heavens as a pious task during the Golden Age of Islamic learning, and the names given to many heavenly bodies, which English speakers can scarcely pronounce, bear witness to this. Copernicus, Galileo, and Newton were men of faith. Gregor Mendel, the father of genetic research, was an Augustinian friar who eventually became the abbot of his order. And it was the traditional Jewish emphasis on education that ultimately resulted in an order of scientists who changed the way we now understand the cosmos. An examination of the interactions of science and religion throughout history, especially in Western Judeo-Christian-Islamic societies, is instructive, reminding us again that the best minds have frequently in the past devoted themselves to science and theology.

Today science and religion are more often felt to be in conflict rather than in harmony. Motion pictures perpetuate stereotypes of both religious obscurantists—raving evangelists with bouffant hair and rattling collection plates—and mad, disheveled scientists screaming "it is alive; it is alive," as they conjure up monsters in their Gothic laboratories. Films like *Inherit the Wind* and *Frankenstein* may entertain, but they do nothing to clarify issues and only caricature both scientists and persons of faith. Feeling themselves frozen out of the mainstream culture, which treats them so shabbily, fundamentalist religious factions have formed their separate society, with creation museums (such as the one in Petersburg, Kentucky), their own presses, and Bible colleges. Scientists also complain that they are misunderstood by an ignorant public and inadequately funded, their contributions to health and prosperity unappreciated.

Biological evolution has become a universally accepted theory in the sciences, and much beneficial research has been founded upon it. Few people, whatever their religious beliefs, would resist necessary medical treatments that rest on this research. While Charles Darwin's evolutionary writings, and the neo-Darwin synthesis that has developed from them, are acknowledged by scientific researchers worldwide, millions of people, especially in the United States as well as in less industrialized countries, still reject the proposition that humans have a common ancestor with other primates and that they are not the result of a separate act of divine creation. To them, the message of Genesis, that humans are made in the image of God, is incompatible with evolution.

Several organizations worked to make theistic beliefs acceptable to a secular, scientific 20th century, hoping to find a meeting ground between the scientific elite and the general public. The American Scientific Affiliation (ASA) was organized in 1941 by the Moody Bible Institute as an ecumenical science organization with Christian evangelistic intent. A movement calling itself scientific creationism formed in the early 1960s, inheriting some of the participants and much of the philosophy of the ASA. Attempting to present in scientific language a belief in six days of creation, Noah's flood, and the special creation of Adam and Eve, the movement gained some popular following among educators. Still, it did not fare well when tested in the U.S. courts, where it was declared an unconstitutional intrusion of a particular religion into the classrooms.

A more sophisticated objection to general evolutionary views has subsequently been put forth by the proponents of intelligent design. The leaders of this movement, for the most part, have scientific degrees from major universities or are distinguished legal scholars. ID, as the movement is known, does not use religious language and does not make its appeals from the Bible. It does not seek to drive evolution out of the schools. It only asks for equal time to present its interpretation of cosmic and human origins alongside evolutionary teachings. But ID has not fared any better in the courts than scientific creationism. Although its advocates have often been skilled debaters who use the specialized languages of the sciences with facility, their opponents, equally skilled at polemics, have accused them of perverting the methods of true science with a thinly disguised religious agenda. With credentialed people on both sides, the controversy has become heated, with name calling and character assassination often part of the discourse.

Issues of separation of church and state are involved in every attempt to introduce evolution, creationism, or intelligent design into public classrooms, along with the ever-increasing difficulty of acknowledging religion in a pluralistic society. No respectable education today can ignore either religion or science. The spiritual and artistic treasures of the great world religions and the developments of modern science are the two great achievements of human civilization. Students should receive the education that best serves them, even when their families cannot afford elite private schools that operate without the restrictions imposed on tax-supported institutions. Sensible, workable definitions of *science, philosophy,* and *religion* need to be clarified. What is the scientific method? What are its strengths and limitations? What is meant when reference is made to a scientific *theory*? Equally, what is a *myth* as the word is often used in religious discussions? Why is myth, within philosophical or religious discourse, something quite different from the popular definition of "wild, preposterous story"? It is further necessary to distinguish between religious indoctrination, which properly takes place in religious institutions and homes, and education about religions, which U.S. courts have encouraged as part of a rich program of secular education necessary for an understanding of art, literature, history, and even current affairs.

This is not a book about science or about religion; rather, it attempts to evaluate dispassionately a central controversy in U.S. society. Americans are often identified as the most religious people in the Western industrialized world. Yet religion has been disestablished since the country's beginnings, and Thomas Jefferson's "wall of separation between church and state" is rigorously maintained by the courts. Because most of the literature of evolutionists, creationists, and intelligent design theorists, abundant though it is, takes a clear, often dogmatic position, this book will attempt to examine the controversy objectively. The heavy reliance on a large number of excellent resources will be quickly evident. Popular science writing is often exemplary, and numerous fine books exist for the general public. The best known writings of the creationist and intelligent design establishments are also designed for the nonspecialist. Readers are encouraged to consult the books especially recommended in the annotated bibliography that completes this work, where fuller presentations of the issues, arguments, and scientific findings are presented.

Because students preparing reports, panel discussions, and term papers will be consulting this book for ideas, the appendix will introduce some of the most intriguing—and often idiosyncratic—people they will ever encounter. Any one of these figures is worth further investigation. Individual chapters will often be consulted by students; for clarity, there will be some necessary repetitions.

A few years ago, I prepared a book on gay marriage for the Greenwood Press series Historical Guides to Controversial Issues in America. At that time, I thought this was the most polarizing issue in the country. Now I know otherwise. No subject arouses more passion than religion. And for many scientists, their discipline is itself a religion. Most university science professors—a group that includes many of my professional colleagues—look frightened, slightly angry, and much on their guard when the subject of intelligent design is mentioned. And some of these folk are even active in their churches. Any mention of a supernatural creative force, they tell me, is inappropriate in a science class and should properly be addressed, if at all, in classes of philosophy and the history of religions. Yet one of the central problems often identified in university education is the compartmentalization of knowledge. Interdisciplinary courses, even when acknowledged to be desirable, are very difficult to construct and are especially fraught with controversy when the sciences are involved, as directors of interdisciplinary honors programs around the country will testify. But teachers of religious subjects are also faced with problems, because the objectivity required in public institutions is difficult for many people to maintain in that they do not understand how religious subjects may be discussed descriptively without any resort to pious sentiment or attempt at moral uplift. Some teachers find the task impossible.

My own interest in the interactions between science and religion started in my public school days in Tennessee. At that time, instruction in evolution was not only considered ungodly; it was actually illegal. No doubt the state universities ignored the law prohibiting the teaching of theories that denied the Biblical account of creation, popularly known as the Butler Bill, which remained on the books in Tennessee until 1967, but my high school teachers remained law abiding and mentioned William Jennings Bryan with reverence. The result was that by the time I entered college, my knowledge of science, such as it was, came from the reading of science fiction, which somehow did make its way into Tennessee libraries and paperback book stands.

We all knew how John Scopes had been convicted in the "Monkey Trial," which took place in Dayton, Tennessee, in 1925. Years later, in the late 1960s, I had the good fortune of meeting John Scopes near the end of his life, when he spoke at George Peabody College in Nashville, where I was an English professor. He was an unassuming man who seemed still a bit bewildered by his celebrity. Scopes and his wife had just returned from Dayton, where he had been given the key to the town and honored in festivities, along with the granddaughter of William Jennings Bryan and other living descendants of the principals in the Scopes trial.

I now own a farm only a few miles from Dayton. It is located in one of the most visually splendid regions of the United States, in the Sequatchie Valley, ringed by mountains. Dayton itself, just as it was in 1925 when it had its moment of fame, is a pleasant mountain town, populated by hospitable people. In the basement of the courthouse where the famous trial took place, a modest museum commemorates the event, and the local motel displays behind its reception desk a mosaic of a scampering monkey. Bryan College, founded after the trial, still trains devout students, and I am told that the biology department is headed by a Yale University PhD whose doctoral work was supervised by Stephen Jay Gould, one of the most famous evolutionary scientists.

I have known several participants in the court cases mentioned in this book, and I have served as an expert witness in church-state trials, though not the ones discussed here. I have also lived much of my life in that part of the country known as the Bible Belt, the region derided by H. L. Mencken during the Scopes trial as "the Bible and Syphilis Belt." For a number of years, I lived in Nashville, sometimes called "the Buckle of the Bible Belt," a center of religious publication. When I see *Inherit the Wind* on stage or screen, I still bristle at the way Tennessee people are caricatured. But I have also felt, from time to time, and not just in the American South, the oppressiveness of religious bigotry.

As a teacher of literature and the humanities for many years, I have complained loudly of the biblical and religious illiteracy that exists even within the South itself. It is, we all concede, impossible to fully understand Western civilization and its arts without some knowledge of the Judeo-Christian scriptures and religious traditions. I have found that students, no matter how active they tell me they are in their churches, often give remarkable answers to questions about the Bible. One student believed the Gospels were written by Matthew, Mark, Luther, and John. Another thought that Jesus lived during the reign of Queen Victoria. Today biblical allusions must be carefully explained to most young people. They do not understand what Martin Luther King, Jr. meant when he prophetically said, shortly before his assassination, "I've been to the mountain, and I've seen the Promised Land." All over the world, wherever there is television, the name "Oprah" immediately conjures up the image of a smiling Oprah Winfrey. Yet how many people, even in the South, would be able to identify her biblical namesake, Orpah, the Moabite sister-in-law of Ruth? Or for that matter, how many graduate students understand John Keats's reference to standing in "the alien corn"?

Several years ago I asked a college class in Nashville to identify Esther, or Hadassah. Nobody recognized, by either her English or Hebrew name, this heroine of the Bible and, according to the scriptures, ancestor of both King David and Jesus Christ. Yet everybody knew Archie Bunker, hero of a television sitcom popular during the 1970s. A few years later, even deeper into the Bible Belt, in Montgomery, Alabama, I was attempting to teach an undergraduate class a short story entitled "Flowering Judas" by Katherine Anne Porter. Although my students were usually alert, they seemed befuddled by this narrative. Fortunately, one student finally raised his hand and asked me who Judas was. It had not occurred to me that I would have to identify Judas Iscariot to a group of Baptists, Methodists, and Pentecostal Christians, but when I explained the allusion, the story, with its intricately woven symbols and theme of betrayal, became clear.

In colonial America, John Milton was more widely read and deeply appreciated than Shakespeare. As late as the early 20th century, one university professor informed his students that the human race had produced two great minds, Jesus Christ and John Milton. No secondary education was complete unless a student had read *Paradise Lost*. Yet today Milton is largely banished from the classroom. Even graduate students in English do not necessarily read Milton, and one English major at a major university, a straight-A student, identified Milton as the author of *The Canterbury Tales*!

Milton's religiosity, which was never orthodox Christianity, is not what keeps him out of the classroom. He is almost impossible to teach today in the time available for a literature class, because he requires of his reader a knowledge of both the Bible and classical mythology. The study of Milton today, when it does occur, has become an exercise in footnoting, with little time left to savor Milton's own "mighty line" and powerful ideas.

A trip to a major art gallery, unless it is devoted totally to modern art, is a shallow experience for someone who knows nothing about classical mythology or the Bible. Even in the old Soviet Union, which was hostile to all religion, children were taught to identify biblical characters and events in order to understand traditional icons on their trips to major art galleries. Christianity and Buddhism have inspired the world's greatest bodies of art, while classic Islamic architecture is another world treasure. Some knowledge of religious history and custom is essential for a fuller appreciation of these masterpieces.

In earlier centuries, homes throughout the American wilderness could be counted on to own two books, the Holy Bible and the Home Medical Encyclopedia. Today, if the Bible is still found in the parlor, it is likely to be covered with dust. Americans once had a common cultural frame of reference; they understood what Abraham Lincoln meant when he spoke of a house divided against itself. As late as his presidency in the 1960s, Lyndon B. Johnson is said to have had two speech writers devoted to Bible quotations, one for the Jewish scriptures and another for the Christian. Today, presidents are more likely to take their quotations from popular motion pictures and television commercials. "Come on, make my day," Ronald Reagan told an opponent, while George Herbert Walker Bush asked, "Where's the beef?"

Not only is the Bible the germinal work of literature of the Western world; it is a part of human culture. The lore of Judaism, through its two daughter religions, Christianity and Islam, has been carried throughout the world. An understanding of current events still requires an awareness of the common heritage of Jew, Christian, and Muslim through the patriarch Abraham.

Because I have been a teacher of the humanities, my lament has been the absence of biblical literacy. But scientific illiteracy is possibly even more pronounced. How many people today are going through life with a medieval understanding of the cosmos, essentially living on a flat earth, because Copernicus, Newton, Darwin, Einstein, and Heisenberg are no more than names they have heard in passing? As C. P. Snow observed, not to have read *War and Peace* is to be educationally deprived, and to have no understanding of the Second Law of Thermodynamics is to be equally disadvantaged. Without some knowledge of science, one lives in a world where things might as well happen by magic. Cyberspace may be familiar territory to young people today, but without scientific knowledge, its workings remain as mysterious as the magic of Narnia.

Clarksville, Tennessee, 2009

1

Science and Religion: A Brief Historical Overview

Today many people assume that science and traditional religions have long been at war. However, historians of science point out that this is most incorrect.[1] Science and religion, often working together, have been the two most potent forces shaping civilization, and their role in human culture is far from ended.

THE TWO BOOKS

More often than not in the past, it was religion that spurred scientific development. From ancient Zoroastrian astrologers who searched the heavens for signs from God, through the Islamic mathematicians, astronomers, and geographers of their Golden Age, to the Western Renaissance and modern era of Copernicus, Galileo, and Newton, devout persons studied the material world and searched the heavens for revelations of the divine.

In the pragmatic applications of their mysteries, scientists and theologians have exhibited similar goals. Though driven by curiosity about the heavens and earth, both have also sought to improve the human lot. Evangelists outlined concrete programs of salvation to their faithful and plans for bettering society through ethical conduct, while scientific researchers sought ways to make life tolerable by feeding the hungry more efficiently, preventing disease, and curing afflictions. Religions have sometimes promised resurrection, survival beyond the usual three score and ten which are the allotted human years. But visionary scientists have even spoken of a time when the burden of sickness and death might be mitigated or lifted through medical breakthroughs. While numerous ancient and contemporary philosophies and religious systems have taught that this life is a realm of necessary suffering based on no more than the illusion of reality, the technology of the Western world, with its enormous benefits as well as its possible demonic abuses, rests on the Judeo-Christian-Islamic acceptance of the genuine reality of the material world that a Creator once pronounced "good."

The common belief that religion works by faith only, while science deals with solid logic, reason, and observation requires some correction. Saint Paul's elegant definition of faith as "the substance of things hoped for, the evidence of things unseen" is often quoted. But Christian theologians, priding themselves on the exercise of reason, have always asserted that true faith is not credulity. Science, too, though it may not always readily admit it, operates through faith statements and aligns itself with particular schools and theories, even as it uses experimentation and observation extensively. While scientific discovery has sometimes started in a burst of inspiration or intuition, usually regarded as the tools of religion, theologians have often made pronouncements about subjects such as anthropology or geology that scientists regard as their territory.

In 1999, Yale professor and scientist Stephen Jay Gould set forth his less-than-original doctrine of nonoverlapping magisteria, which he baptized NOMA. He felt that science and religion could coexist peacefully if each recognized its own proper sphere, without overstepping proper bounds into the province of the other. Science, he claimed, is in charge of the natural world, while religion's proper concern is morality and ethics. Although Gould presented his position with considerable charm in his book *Rock of Ages,* in reality science and religion have never been two separate areas addressing separate questions and reaching separate conclusions. Each is always invading the Gould-appointed province of the other.

The first five books of the Hebrew Bible, traditionally known as the Five Books of Moses, set forth an account of the world's beginnings. Though these books have been widely attributed to Moses himself, they recount his death. (After all, he was a prophet!) "In the beginning God created the heavens and the earth" commences the biblical narrative, which moves on to an account of God's subsequent disillusionment with his creation and his determination to destroy it. He relents by choosing Noah and his family, along with pairs of beasts, for survival, while the rest of the world is destroyed and purified with water. The biblical narrative continues with the selection of Abraham to be the father of a prophetic nation, the Israelites. Later, Moses, an Egyptianized Hebrew, leads the Israelites out of Egyptian bondage into the desert, where he receives the Ten Commandments from God. Although Moses died before achieving the Promised Land, the Israelites were led by Joshua to a conquest of Palestine, where they built a temple to God and made Jerusalem their holy city. In the Christian Bible, or New Testament, the divine plan is continued, with the Incarnation of Jesus, a man widely conceded to be historical and believed by Christians to be also God incarnate. Through the life and crucifixion of Jesus, it is believed, the promises of the Hebrew deity are extended to all humankind. The scriptures conclude with a prophecy of the end of the ages.

Even today, millions of people accept these biblical narratives literally. Some spend years scrutinizing the books of Daniel and Revelation, attempting to discern the "end times," freely identifying among national leaders such figures as "the beast" and "the anti-Christ." Others have observed that the books of the Bible fall into a number of literary forms well known in the ancient world, and the conventions of different genres must be considered in any valid interpretation of them. It is further suggested that these narratives, like others of the ancient world, are characterized by tales of the miraculous intermingling with historical facts and figures and are fashioned to the understanding of the people who first received them, people as yet unenlightened by Galileo, Newton, and Einstein.

Even secular scholars have acknowledged that the ancient Hebrew religion has bestowed upon the world concrete benefits. The concept of a Sabbath, one day in seven free of labor for both humans and beasts, is one of the great contributions to health and welfare now observed in almost all parts of the world. While modern historical research suggests that the observance of a Sabbath was not necessarily unique to the ancient Hebrews and may in fact have been borrowed by them from another ancient society, they are certainly the ones who have spread the practice throughout the world. And their scriptures anchor the Sabbath in a splendid narrative. For six days, according to Genesis, God labored. On the seventh day, he ceased the labor of creation, consecrating the day and giving it as a blessing to humanity. This narrative is sometimes understood as a myth in the sense of a poetic embodiment of truth, because God, according to Hebrew understanding, neither labors nor sleeps as do humans. Other practices of the Hebrews—such as the Jubilee Year, in which serfs were liberated, and ways of nourishing and preserving the soil that feeds the people—have taught humankind valuable lessons. The Hebrew concern for all living beings—for the poor, and for widows and orphans—was almost unique in its quality. In an ancient world where slavery was universal and animals were often considered no more than beasts of burden, these rules were exceptional.

Aristotle, who lived in the fourth century B.C., may or may not have been a theist as the word is usually understood. The data remain ambiguous. He is usually, however, thought of as the father of Western science, and he does appear to have discerned a clear purpose evident in nature. Although we do not observe any agent deliberating, Aristotle appears to have believed that we are not totally at the mercy of random chaos. His model of the universe, based on the knowledge available to him, was geocentric. The earth, as human senses seemed to confirm, was the center of everything—and it was probably eternal—with the sun, moon, and planets rotating around it.

Other foundational Greek thinkers, before and after Aristotle, postulated a Divine Being, the One. Plato taught that truth exists, even if humans are unable to perceive it. The philosophers—in contrast to Greek poets, who made abundant use of polytheistic mythological lore—believed that a Supreme Being exists, unchanging throughout eternity, totally unmoved by the actions and desires of humans. Later, Judaism and early Christianity, while greatly influenced by the Greek philosophers, affirmed a God intimately involved with humanity. Jews taught that God had chosen a particular people to be a light unto all humankind, while Christians made the radical assertion that God had assumed human flesh. Both Judaism and Christianity taught that history is a stage on which the divine plan is enacted.

In former times, scientists frequently made use of sacred texts. Why do the stars, after all, have the names of ancient deities? Why, when looking up to the southern sky, does one navigate by what is designated as the Southern Cross? Pious thinkers of antiquity, whether Jewish, Christian, or Muslim, held that there were two sources of knowledge about God: the Book of Scripture and the Book of Nature.

This is nowhere illustrated more convincingly than in the thinking of Saint Augustine (A.D. 354–430). Born to a pagan father and a Christian mother, Augustine at first found Christianity illogical, the Bible full of "contradictions and nonsense." Even after his adult conversion to Christianity from paganism, he was unable to reject the wisdom of Greek philosophy, working out a synthesis instead. He constructed one of the first great systems of Christian theology, which continues to influence both Roman Catholics and Protestants today. Augustine accepted the most enlightened science of his time and taught that all truth is one. The seeming contradictions between Christian doctrine and natural philosophy he resolved to reconcile through the use of reason. Accepting God as the author of the Book of Scripture and the Book of Nature, Augustine concluded that they could not disagree. Yet both books required thoughtful interpretation. He taught that biblical passages have layered meanings; in addition to the literal sense, there were allegorical, anagogical, and moral levels of meaning. In this approach to scripture, Augustine owed much to Alexandrian Jewish philosophers who preceded him and followed a similar method of biblical interpretation. Even the enlightened Greek pagans had frequently interpreted Homer and their other literary classics as allegorical. Only readers of the modern age have developed the habit of looking for a single literal meaning in important literature.

For Augustine, comprehension of both scripture and the natural world required the careful use of reason. Interpretations of scripture are always partial, he believed, with new knowledge always expanding understanding. Since scripture was given to people of all ages, the Divine Author had revealed the Word to different degrees of understanding, yet with a profundity that would gradually unfold. Because human languages, ambiguous as they often are, were used in divine revelation, many obscurities waited to be clarified. And interpretations of scripture, Augustine believed, should always be brought into harmony with the current state of knowledge of the material world.

Augustine understood that if Christians failed to conform their teachings to current knowledge of the natural world, they would open themselves to ridicule. While faith should always remain primary, science, as he understood it, should serve as a helpmeet in the search for true comprehension of religion. Reason and faith constantly correct one another, while the natural world as well as scripture reveal the majesty of the Creator.

Although he was a man of the ancient world, many of Augustine's writings are surprisingly modern. His teachings about faith and reason have been recently echoed by Pope John Paul II. In the encyclical *Fides et ratio,* the pope identified faith and reason as "two wings on which the human spirit rises to the contemplation of truth." Faith without reason, the pope acknowledged, can easily degenerate into superstition. Blind faith and the unenlightened study of scripture alone do not lead to knowledge. Yet without faith, humans are left to wander in darkness. Unaided reason alone, the church had always taught, is insufficient. The pope harshly denounced radical relativism, nihilism, and scientism (the substitute of science for religious loyalty) and concluded that scientific studies, unguided by faith, can be destructive. Perhaps he was thinking specifically of weapons of mass destruction, genetic manipulation, and other applications of modern science.[2]

Throughout the Middle Ages, brilliant theological philosophers appeared in Judaism, Christianity, and Islam. In the West, Saint Thomas Aquinas in his *Summa Theologia* synthesized Aristotelianism and Christianity, providing the foundation for Roman Catholic theology until the present age. Nature, he believed, could not be fully understood without an acknowledgement of divine purpose and direction. Subsequently, religious thinkers continued to speak of the two sources of knowledge, the Book of Nature and the Book of Inspiration.

Almost everyone affirmed the existence of a Creator; yet the degree of direct action of this Creator in the governing of the universe became a matter of discussion and debate. Though we inhabit a cause-effect universe, is a particular causation natural, supernatural, or a mixture of both? How can one know? What is Providence? Can the Judeo-Christian-Islamic faith in a loving, active God be reconciled to the Greek philosophical concept of a being who is omnipotent, omnipresent, omniscient, eternal, and unchanging? Science, which relies on dependable laws, would seem undermined by a deity operating by whims, playing favorites among humans, or violating his own laws when so inclined. A God who plays dice with creation, to use the Einsteinian metaphor, would appear to make science impossible. An important affirmation of 13th-century theologians—Jewish, Christian, and Islamic—is that God can do anything he chooses. However, he restricts himself—almost always—to working within the natural laws he has created. On a popular level, folk religion still tended to explain every event it did not understand as the work of angels or demons.

A few medieval thinkers, anticipating 18th-century Deists, reduced divine activity to an initial instant of creation ex nihilo (out of nothing). After this, they contended, the natural laws took over. In the School of Chartres in France, theologians explained miracles such as the parting of the Red Sea in naturalistic ways, as later in the 17th century the Jewish philosopher Spinoza was to do. Elaborate narratives were constructed; perhaps the winds, the season, and the water all conspired to part the sea just at the moment the Hebrews needed to cross, while the same forces cooperated in swallowing the pursuing Egyptians when they tried to cross. God, in his infinite foreknowledge, had planned this at the instant of creation.

Still, no matter how rational they attempted to be, most religious thinkers left some room for miracles, those events that defied natural laws. After all, the Incarnation and Resurrection, certainly miraculous, were at the heart of the Christian faith. Still, God, they believed, usually worked through natural laws. Jesus was indeed a historical person, born of a human mother. He had an infancy and a youth in which "He grew in favor with God and man." He did not suddenly manifest himself full grown in the manner of a Greek god or goddess disguised in mortal form to accomplish a single feat or consummate an amorous impulse.

The discernment of miracles was always a problem. Were miracles simply events for which natural causes remained obscure? Would advances in knowledge and a fuller understanding of those natural laws eventually put an end belief in the miraculous? Numerous religious folk refused to relinquish their faith in the miracles related in the Hebrew and Christian Bibles, even while they observed that such extraordinary events no longer transpired. In response, Anglican divines and some Protestants developed a doctrine of the cessation of miracles after the Apostolic Age. This doctrine ran counter to the Catholic belief that miracles continue to the present day and still serve their ancient function of aiding the faithful and verifying the faith.

The belief that miracles have ceased led to a reexamination of the function of prayer and even, if taken to its ultimate conclusion, encouraged a brand of Deism that removed an active role of God from the present-day world. Though Roman Catholics continued to affirm that miracles can occur, after the 18th century, even their rationalists found them embarrassing. Consequently, in Catholic practice, rigid tests had to be passed before miracles could be officially confirmed by the church.

GALILEO

Of course, scientists and theologians did not always work in the harmony that Augustine or Aquinas envisioned. Even among the faithful sons of the church, challenges to venerable beliefs emerged. The case of Galileo is often presented as an example of the church's antagonism to science. But the received story that has made its way into many textbooks is legend more than fact. All schoolchildren learn how Galileo, when brought before the Inquisition, wisely denounced the Copernican theory that the earth moves around the sun, which he had earlier accepted, yet whispered under his breath, "but it moves all the same." In reality, the Galileo story is more complex than this simplistic rendering. He believed that scripture must be interpreted in light of current scientific knowledge, just as Augustine and Aquinas had taught. However, he lacked political finesse and had offended the pope and other church dignitaries with his arrogance and sarcasm. In 1633, he did get into trouble because of the tone of his book *Dialogue on the Two Chief World Systems,* but personal conflicts were largely responsible for his problems and his subsequent house arrest during the last part of his life.[3]

Galileo was officially vindicated by his church only in 1979, when Pope John Paul II authorized a commission to examine "the Galileo affair." The church was certainly tired of all the negative attention it had long received over the issue. The papal commission rehabilitated the great scientist, admitted that the officials of the church had committed errors, and reaffirmed Augustine's principles of biblical interpretation, along with the compatibility of informed faith with true reason, whatever these are! Since 1979, there have been few conflicts between Roman Catholicism and pure science, though disputes abound over a number of the applications of modern science. While some individual Roman Catholics have been prominent in the intelligent design movement, church officials have paid it less attention and even occasionally opposed it.

THE AGE OF REASON

Despite the important discoveries of astronomy and new theories of cosmology of the 16th and 17th centuries, the latter century was a period of interreligious turmoil in Europe, with less energy devoted to matters scientific on the part of many churchmen. Still, as the result of scientific inquiry and other challenges to the traditional faith, the metaphor of the universe as a gigantic functioning clockwork became common among intellectuals. With each scientific advance, the Deity seemed less necessary to the daily operations of the world. For many intellectuals, only a "God of the gaps" seemed to remain, retreating constantly as science offered more and more explanations of natural events that had earlier appeared mysterious or miraculous. Such a God did not appear to be a very significant force. Even if the gaps might never be totally closed by science, theology appeared to have become a defensive operation.

According to the Deists, becoming more vocal in the late 17th and early 18th centuries, God was an absentee landlord who, after setting the universe into motion, no longer chose to interfere with its maintenance. Did this make spirituality merely a delusion leading eventually to atheism, with dire consequences for the masses of people who would no longer have an omnipotent Being to demand an explanation for their conduct? It is not surprising that some French social theorists concluded that if God did not exist, he would have to be invented.

There were, nevertheless, three scientific thinkers of the 17th century who turned their thoughts frequently to religious questions. Robert Boyle (1627–1691) was a deeply religious man who considered the natural philosopher (his term for scientist) a "priest of nature." He regarded his own study of nature, much in the manner of a medieval scholar, as a religious exercise. He interested himself in alchemy, the precursor of modern chemistry, believing that a "philosopher's stone," when discovered, would draw angelic beings to itself.

The second religious scientist of the 17th century was the great Sir Isaac Newton (1642–1717). One of humanity's prime geniuses, he is still regarded as among the most creative scientists of all time. Though not in the orthodox Anglican tradition, he was a very religious man. In pondering the Christian scriptures, which was frequently his custom, he tried to correct what he believed to be "corruptions," defined as places where the Bible contradicted his own conclusions. In his *Principia Mathematica* (1687), one of the basic

Western scientific treatises, he discussed the "attributes and activities of God" in ways that would be totally unacceptable in scientific writings today, which disallow appeals to the supernatural. Newton wanted to locate evidence of continued divine activity in the world, and he may have regarded gravity itself as such evidence. When irregularities in the observed movement of the planets appeared to defy his laws of gravity, he was ready to speculate that God directly intervened to offer a "correction."

In 18th-century Europe, many intellectuals revolted against the violence of competing religious sects that had characterized the previous century. For them, a new universe revealed by Copernicus, Galileo, Newton, and other scientists required a revision of religion. They believed in examining all inherited beliefs with their new tools: empirical evidence and historical investigation. Many who chose to anchor their beliefs in historical Christianity recognized that it was a religion that had changed and expanded during the centuries. While most intellectuals still looked at civilization as a work in progress, with humans moving ahead at each step from ignorance into enlightenment, there were those who felt that the age of truth lay in the past. These were the restorationists, who believed that a study of the past would provide proper answers to challenges of the present. Others resorted to pietism, grounding their faith in illumination and enthusiasm, in mystical approaches to religion that could not be disproved by science.

But the religious response felt to be most representative of the Age of Reason was still Deism, a religious philosophy attracting artists and intellectuals, especially in France, England, and North America. Deism made a cult of reason and believed that it could supply all needs. While a belief in divine revelation might be essential for the masses of people, sustaining them in hope and protecting them from moral chaos, reason alone was sufficient for thinking people. Since a nominal religious affiliation frequently was held by Deists, some of whom even penetrated the hierarchy of the Anglican Church, others joined emerging groups such as the Congregationalists, Universalists, and early Unitarians. It was popular among Deists, as among many in the free churches, to attack "priestcraft," meaning the abuses, real or alleged, of the established churches. A common belief among them was that Jesus had been a gentle, pastoral teacher of good manners and good conduct but that his life and teachings had been contorted by religious officials.

With a Creator now removed from his creation, perhaps watching it from afar, "paring his fingernails" as one novelist has expressed it, philosophers elevated reason above any notions of divine inspiration. Yet reason still seemed to dictate to them that there was a Creator, who had perhaps constructed "the best of all possible worlds," where whatever exists is right. Earthquakes killing thousands and other natural disasters, which we still refer to as "acts of God," did not sway some from this belief. Voltaire's witty *Candide* (1759) was both a work of enlightenment and a critique of its ideas, especially the belief that we are living in the best of all possible worlds. It has been said that, on his deathbed, Voltaire, like many free thinkers, called for a priest to give the last rites, yet got into a heated argument with the priest as he lay dying! Though the story may be apocryphal, it illustrates the ambivalence of the age.

A good example of a Deist (though he is perhaps more accurately defined as a Unitarian) was Thomas Jefferson, the American founding father who fashioned the Declaration of Independence and is often thought of as the architect of the nation's church/state policy. A free thinker, he found the Bible a useful and often inspiring (as distinguished from *inspired*) book. Among his writings is "The Jefferson Bible," a result of contemplation and study of the life of Jesus. Jefferson wanted to rescue Jesus from the two forces he believed had distorted his message: the evangelists and the established churches. To Jefferson, Jesus was a wise, eloquent, and noble teacher done to death by the Romans and then grossly misrepresented by the churches. In compiling his bible, Jefferson used the scholarly tools available to him: Greek, Latin, French, and English texts. He thought he could separate biographical facts and Jesus' authentic teachings from the miraculous framework he believed had been superimposed on them. Accounts of miracles and mystical sayings were removed from his book, as Jefferson unintentionally projected his own sensibility, including his personal strengths and weaknesses, on the founder of Christianity. Jefferson's fallacy, of course, was that his only sources of the life and teachings of Jesus were precisely those he had already declared unreliable—the scriptures themselves.

The most important American theologian of the 18th century was Jonathan Edwards (1703–1758). Though he regarded himself a faithful Christian, he fully embraced the Calvinist doctrine of predestination, which in many ways made God as currently inactive as the Deists believed. Even the salvation or damnation of each human had been determined before that person's birth. Edwards's eloquent and learned sermons combined Calvinist theology with ideas drawn from natural theology. Though he valued sense perception, he also believed that through intuition humans could reach some perception of God.

Representing a revolt against the cold intellectualism of the Deists and predestinarian Calvinists, Ralph Waldo Emerson (1803–1882) was widely regarded as a wise man who brought enlightened religious ideas to many Americans and Europeans. Influenced by Asian spiritual ideas and the new European movement known as Romanticism, he rejected the belief that reason alone is the primary or only path to truth. Emerson, in endlessly aphoristic sentences, taught that the world is the product of one mind and will, which is active everywhere. By this guiding principle, humans are safe, and the universe is comfortable and habitable. (This might be considered a variation on the Taoist theme of "man's at homeness in the universe," or, in modern scientific jargon, "the anthropic principle.") For Emerson, sin was not a meaningful concept, and he did not dwell on evil or pain. Much of his thought, modified and divested of his own language, lingers on in popular religion, especially that of the "feel good" television evangelists of the 21st century.

DARWINISM

From the middle of the 18th century to the middle of the 19th, European naturalists and biologists were busy presenting theories and findings that were new and troubling to the general population. Even before Charles Darwin's 1859 *On the Origin of Species,* the Comte de Buffon, Jean-Baptiste Lamarck, and Charles Darwin's own grandfather, Erasmus Darwin, were suggesting that animal species were more closely related than had before been considered and that they might actually change through many millennia.

An important figure in the history of popular religious thought and its relationship to science is William Paley, whose book *Natural Theology* was published in 1802. The book was required reading in seminaries in both England and the United States. As a divinity student, young Charles Darwin studied it faithfully. Paley used natural theology arguments from former times, giving them a new liveliness. His most famous image of a watch found on a forest path was to be used dramatically by pulpit preachers for over 150 years. This watch, he said, would be immediately recognized as an object carefully fashioned by a watchmaker. Likewise, the world, with its pattern and equilibrium, had to be the work of a Maker. In the 20th century, Sir Fred Hoyle would bring the argument up to date when he compared the chance-assembled universe of atheistic science to a Boeing 747 randomly constructed from trash by a tornado blowing through a junkyard.

Devoutly religious people had so far managed to reconcile their piety with emerging scientific knowledge. But the greatest challenge to conventional religious belief came with the publication of Charles Darwin's *On the Origin of Species by Natural Selection,* with the confirming views of Alfred Russel Wallace, who had independently reached basically the same conclusions. Though the subsequent paths of the two scientists would diverge—Darwin would become agnostic, while Wallace would give himself over in later years to spiritualism—they were alike in their conclusions that species were not individually created but had changed over time in response to pressures of natural selection and the struggle for existence. Darwin's original book gingerly dealt with religious implications; he obviously did not want to be thought an enemy of faith or to offend his pious wife. It was his *The Descent of Man* (1871) that really shook the public. Many were willing to concede that animal and plant species might change—farming practices had already given hints of this—but they were unwilling to surrender their belief that humankind was a special creation, in the image of the Creator, rather than one of several primates with a lowly common ancestor.

Darwin's publications attracted the controversy in England that he feared. In 1860, a celebrated debate was held at Oxford University. The esteemed Anglican bishop Samuel Wilberforce attacked Darwin's ideas at a meeting of the British Association for the Advancement of Science. His opponents were two scientists, Joseph Hooker and the more famous Thomas Henry Huxley, who would subsequently become known as "Darwin's bulldog" as he continued his defense of Darwinism throughout his life. The debate at moments became personal, as when Bishop Wilberforce is said to have asked Huxley whether he was descended from apes on his mother's or his father's side. Huxley is supposed to have responded that he preferred descent from an honest animal than from a learned clergyman who used his intellect to ridicule what he could not understand. Both sides claimed victory in the debate, and even today their arguments are repeated in intelligent design confrontations. Huxley's book *Evidence on Man's Place in Nature* (1863) was an important contribution to the controversy, and the next two generations of his family, with their eminence in both science and literature, carried on the tradition.

Though the beginning of the 20th century saw Darwinism in decline, the emergence of a neo-Darwinian synthesis, with genetics, paleontology, and other sciences offering further support, brought vindication. Despite earlier scientific as well as religious opposition, Darwin's ideas now gained wide acceptance. The work of Albert Einstein and other physicists would give further authority to Darwinism, with a picture of the entire universe that is constantly changing and in movement rather than static in space and time. Many religious thinkers, feeling that the evolutionists had won the arguments, set about reconciling their theology to what had become established science.

REACTION

But the opponents of Darwin's ideas were not quiet. In the United States, where people were deeply invested in traditional religion and not eager to make any concessions, opposition became more vocal than ever. In 1923, George McCready Price published his widely received book *New Geology*. Price, a devout literal student of the Bible, was largely self-trained in science. As a Seventh-Day Adventist, he had the additional task of reconciling his science to the visions of the denomination's founder, Ellen Gould White, who believed she had been vouchsafed a personal vision of the seven days of Creation. Price was a convincing speaker who presented his ideas with passion, and he soon had a large following within and outside his denomination. He would even be referred to as an authority by William Jennings Bryan during the Scopes trial.

Price taught what would become known as flood geology, that earth's geological features, identified by Charles Lyell and other geologists were the result of Noah's flood, which had covered the entire world. All the sedimentary layers of the earth, complete with their fossil deposits, were, he explained, the result of this upheaval, revealing organic life that had existed before the flood but had not survived.

Price did not attempt to answer all the questions that such a theory called forth. For example, he did not explain how pairs of all the animal species known in the early 20th century could fit into an ark, the specifications of which are provided in scripture. He did not explain how Noah provided meals for all his human and animal passengers or how he kept predatory animals from making meals of each other. One of Price's readers has suggested, no doubt facetiously, that Noah was able to shrink-dry the animals; others proposed a hypnotic state of suspended animation. Neither Price nor the devout British zoologist and Pilgrim Brethren preacher Phillip Henry Goss, despite accusations by their opponents, believed that God had planted the fossils in rocks to tempt infidel scientists. And it was Mark Twain, rather than Price, who irreverently wrote in *Letters from the Earth* (published posthumously in 1962) that God compelled Noah to sail back and get the flea, which he had initially left out of the ark.

The 1920s were a crucial time in the United States, with a conflict between religion and Darwinism boiling. William Jennings Bryan, perhaps the most powerful orator in U.S. history, had already distinguished himself in politics and government service before becoming the most vocal opponent of Darwinism. He believed that harmful eugenics policies had resulted from Darwin's theories, which had also formed a philosophical basis for German militarism, resulting in the Great War. Furthermore, he believed that parents had a right to determine what their children were taught in the public schools financed by their taxes. Always the defender of the common person, he felt a scientific elitism was creeping into education, challenging religion and violating the wishes of the vast majority of parents and citizens. With all this in mind, he accepted the challenge of the Scopes trial in Dayton, Tennessee. The Butler Bill, passed by the Tennessee legislature in 1925, which Bryan defended, survived the trial and remained on the books in Tennessee until 1967, still legally prohibiting the teaching in public schools of any theory of human origins that contradicted the Bible.[4]

Dissatisfaction with the way religion was treated in the public schools was in the air. Beginning around 1948 and continuing over the next decades, several cases reached the U.S. Supreme Court. Religion classes and ceremonies were gradually banned from public schools, until these institutions became totally secular. Even as the courts remained generally hostile, various proposals were entertained to allow some semblance of religion back into the schools, which were still educating the children of highly religious people. Nondenominational prayers and Bible readings without comment were the practice in many schools. Once, *nondenominational* had referred simply to a prayer that did not offend any Protestant church. But as communities became more and more pluralistic, it was ever more difficult to determine exactly what sort of prayer would be acceptable. Even moments of silence at the beginning of school days became controversial, and Bible readings were equally divisive. Jewish, Catholic, and Protestant Bibles differed, often in minor ways that each group still considered important, and Bibles used in schools had to be free of the marginal notes that were in some editions highly interpretive. Released time for religious instruction, either on or off campus, was instigated in some school districts, always against protests. Even the posting of the Ten Commandments in school buildings resulted in heated court cases. Although classes on the Bible as literature or world religions were still legal if part of a program of secular education, it was difficult to maintain the objectivity of such programs or of the people who taught them.

Roman Catholics maintained a vast system of parochial schools, distrusting the public schools as bastions of Protestantism. Orthodox Jews and Lutherans also preferred their own schools when possible. Yet those who supported religious schools were also taxpayers whose money financed the public schools, and they understandably complained of what amounted to double taxation. With the decline in religious vocations and fewer brothers, nuns, and priests available as teachers, it became increasingly expensive for Roman Catholics to staff their schools. In the latter part of the 20th century, the home school movement, chiefly among evangelical Protestants, was largely, though not entirely, fueled by the fear that the public schools had become secular, amoral, and violent. The quality of home schooling varied enormously, and children who were educated at home were deprived of important experiences as citizens in the vast melting pot, where people of many social classes and ethnicities had earlier been Americanized.

Even as the educational system and the larger society were trying to come to terms with evolution, additional scientific developments, emerging with such rapidity throughout the 20th century, raised more questions. In 1953, Stanley Miller and Harold C. Urey attempted to construct a laboratory environment conducive to the emergence of life from inorganic matter, much as they assumed it had happened at the beginning of life on earth. Though their experiment was not immediately successful, it raised hopes of advances to come. Also in that year, the structure of DNA was revealed in the publications of Francis Crick and James Watson. It was not long before modest cloning would prove possible. Stem cell research and new possibilities of genetic tampering would raise further ethical and religious questions. The courts would soon be called upon to issue decisions that would have perplexed Solomon himself. Three women would appear in one court claiming parentage of the same child. With genetic, gestational, and adoptive mothers all possible, the courts would soon have to identify the legitimate parent. Society faced anew the old fear of Faustus, of Frankenstein, of the mad scientist who goes too far in challenging the prerogatives of God.

Meanwhile, the religious proponents of Darwinism were not idle, though they faced new challenges. In the late 1950s, the rivalry between the United States and the Soviet Union led to a demand for better scientific education, especially after the Soviets sent a man into space. More money was channeled into scientific research, and textbooks, which had been reticent about evolution and other controversial topics, were updated and made more scientifically explicit. Not surprisingly, this resulted in a backlash from those who had not renounced their reliance on sacred scripture in favor of science. In fact, the fear of "godless communism" made them cling to religion with renewed fervor.

The movement known as creation science may be said to have started in 1961 with the publication of *The Genesis Flood* by John C. Whitcomb, Jr., and Henry M. Morris. Instead of relying on the two books of nature and revelation, in the manner of ancient scientists, the plan now was to use science to affirm the truth of religious teachings. With more religious people being trained in scientific disciplines, it should now be possible, or so it was reasoned, to clothe religious affirmations in the language of science. But the creation science movement, despite its wide following and after a few successes in states with strong Fundamentalist constituencies, was rejected by mainstream science and did not fare well in the higher courts. Most scientists and the courts alike recognized the movement as religious rather than scientific and appealed to constitutional First Amendment protections against sectarian incursions into the public schools. In 1968, anti-evolution laws were struck down in Arkansas, just as the Butler Act in Tennessee had been finally repealed the previous year.

With creation science largely discredited and driven out of the schools by the courts, people who still found problems with evolutionary theories, more of them than ever with advanced science degrees, took another path. In 1973, an astrophysicist named Brandon Carter popularized the "anthropic principle," which acknowledges that even minor changes in the environment of earth and the unfolding of the universe would have made human life impossible. The odds against human life evolving had been astronomical, yet humans existed, were thriving and multiplying. For religious people, this seemed added evidence of design by a higher intelligence. Likewise, after Fred Hoyle named it and mass circulation magazines popularized the Big Bang theory, it was widely accepted as yet another verification of the traditional Christian doctrine of creation ex nihilo. *The Mystery of Life's Origins: Reassessing Current Theories,* by Charles B. Thaxton, appeared in 1984 and attracted a wide, appreciative following. Thaxton's book is often identified as the first important work of the intelligent design (ID) movement.

INTELLIGENT DESIGN

Intelligent design differed from earlier creationist movements in that it did not attempt to drive Darwin out of the classroom. In fact, some scientists within the movement accepted many features of Darwinism. What ID requested was equal time in the classroom, where students would be exposed to Darwinian and neo-Darwinian ideas but would also be told of the gaps and limitations of evolutionary theories. Students would be presented arguments in favor of intelligent design in the universe, but in scientific rather than religious language. Because different theories of origin would be presented, students would be allowed an important exercise in critical reasoning.

A textbook sanctioned by the ID movement, *Of Pandas and People: The Central Question of Biological Origins,* was published in 1989 and promoted as an appropriate text for introducing students to ID. Though the work of two scholars with scientific credentials, Percival Davis and Dean Kenyon, the book was widely criticized by the scientific establishment, its errors and limitations well publicized. It was dismissed by them as a work of religion disguised as science.

Two years after *Of Pandas and People* was published, a group of similarly minded scholars and teachers founded the Discovery Institute in Seattle, Washington. This has remained the central think tank of the movement, providing speakers and publications clarifying the aims and methods of ID. Perhaps the most influential nonscientist that the movement attracted has been Phillip E. Johnson, a distinguished professor of law at the University of California, Berkeley. In youth, Johnson clerked at the Supreme Court for Chief Justice Earl Warren, and he later pursued an extraordinary career in jurisprudence and legal education. One of the most respected legal scholars in the country, Johnson concluded, after carefully reviewing the major writings on evolution, that the theory rested on so little evidence that it would never hold up in a court of law. His own book, *Darwin on Trial,* came out in 1991, at the same time the courts were viewing ID skeptically. The book gave an enormous boost to the movement, providing a new dimension to its argument. Gallup polls demonstrated that the American people were favorably inclined to ID, with 90 percent of people surveyed affirming their belief in a world created by God. A substantial number of respondents also said they believed the work of creation took place in six earthly days, exactly as the Bible related.

Several other influential books followed, written by scholars with advanced scientific degrees: *Darwin's Black Box,* by Michael Behe, appeared in 1996, and William Dembski's *The Design Inference: Eliminating Chance through Small Probabilities* was published two years later. The movement would have its brief moment of academic recognition when Dembski, in 1999, became head of Baylor University's Center for the Study of Intelligent Design. Though the president of the university was supportive, believing that an institution with historic Christian connections should affirm the faith, there was such strong objection from the science departments and others in the university that the center was closed five years later.

ID advocates have been thrown back on their own institutions, with the Discovery Institute Center for the Renewal of Science and Culture, located in Seattle, Washington, taking a lead. It was here in 1992 that the Wedge Document which has been much commended and much maligned, took form, enunciating goals for the ID program and suggesting techniques for facilitating them. The Wedge Document outlined plans for research in paleontology and molecular biology, with results to be publicized in books, conferences, seminars, and teacher training programs and on public television programs. The stated goal was to defeat scientific materialism, with its "destructive moral, cultural and political legacies," and to replace it with a theistic understanding of human existence.

Many within the scientific establishment considered the Wedge Document subversive, a dangerous threat to the integrity of U.S. science and science education. The scientific establishment received a strong endorsement in 1996 when Pope John Paul II made clear his belief that evolution is now "more than a hypothesis" and that there is no "essential contradiction between evolutionary science and Catholicism." This seemed to end the precarious relationship between Catholicism and evolution that had caused such problems for the Jesuit paleontologist and philosopher Teilhard de Chardin. Catholic scientists were now free to base their work, without further qualification, upon Darwinian assumptions. Catholics still, however, differed from many modern scientists in their assertion that God is indeed the source of life and matter and that he still intervenes from time to time in his creation.

By the end of the 20th century, the Zogby organization and other opinion pollsters were discovering that, despite everything, most Americans favored the teaching of arguments against evolution along with proofs of its validity. U.S. Senator Rick Santorum, a Republican of Pennsylvania, with the assistance of Phillip E. Johnson, proposed an amendment to the popular No Child Left Behind Act stipulating the teaching of intelligent design along with the latest scientific theories and discoveries. Though the amendment was later removed from the bill, it did occasion much discussion favorable to the aims of the Discovery Institute.

Several states became involved in the intelligent design controversy. Many parents and citizens, and some educators, argued for the presentation of two points of view—ID and evolution—citing the need for values clarification and critical thinking in education. Opponents still protested that ID would bring thinly disguised religious instruction back into the schools. The issues played out in the courts of several states, and ID again was strongly on the defensive with science educators.

With many Americans dissatisfied with the courts as well as their schools, there were several interesting developments in 2004. A CBS News poll found that half of Americans contacted still said they believed God had created humans in their present form. A Gallup poll conducted about the same time revealed that, of those polled who accepted evolution, only a small percentage believed it to be a totally natural process, while over three times as many believed that God had guided the process at every point.

One complaint of scientists affiliated with the ID movement was that their work was ignored by peer-reviewed scientific publications. Stephen Meyer, a director of the Discovery Institute's Center for Science and Culture, had the first scientific paper from the movement published in an unaffiliated peer-reviewed journal, *Proceedings of the Biological Society of Washington*. Though this was regarded as an advance by the ID community, the editor of the publication was later severely castigated by his scientific colleagues. Thus, the belief of ID proponents seemed confirmed that academia and the scientific research communities were conspiring to exclude them, their views, and their research from any serious consideration. For the Discovery Institute, this had become a matter of academic freedom.

The new millennium saw a great cultural divide in the United States, and ID was a significant issue. Attitudes toward politics, social mores, economics, war and peace, and capital punishment were all part of a looming crisis. Thoughtful people recognized that when large numbers of the most responsible citizens lose confidence in their basic institutions—particularly their schools—and retreat into their own enclaves, democratic society itself is challenged.

NOTES

1. Although there are several excellent histories of science, this chapter is particularly indebted to the brilliant lectures of Lawrence M. Principe for The Teaching Company, 2006. This series of lectures in philosophy and intellectual history is titled "Science and Religion," and may be purchased in CD and DVD formats from The Teaching Company, Chantilly, Virginia.

2. John Paul II, *Fides et ratio.* http://www.vatican.va/holy-father/john_paul_ii/en cyclical/documents/hf_ii_enc_14091998.fides-et-ratio_cn.html. See also John Paul II, "Address to Pontifical Academy of Science," October 22, 1996. www.ncseweb.org/re sources/articles/8712_message_from_the_pope_1996-1-3-2001.asp.

3. According to Lawrence M. Principe, the key issue in the Galileo affair was the split between realist and instrumentalist views of science (see Lecture Six, "Galileo's Trial," of his Teaching Company course). The realists believed that scientific theories are actual depictions of the world, while the instrumentalists held that these theories are simply tools for structuring explanations of phenomena. Copernicus, Kepler, and Galileo—like most scientists today—were realists, while the church authorities and traditional astronomers of that time were instrumentalists. Each position, according to Principe, rests on a faith affirmation.

4. There is a wealth of published material on the Scopes trial, but the best sources are the books of Edward J. Larson. His lectures for The Teaching Company, titled "The Theory of Evolution: A History of Controversy" are also highly recommended. See Bibliography.

2

Charles Darwin and the Darwinian Revolution

At the beginning of the 21st century, magazines and newspapers had contests to determine the "person of the millennium." One newsmagazine featured Martin Luther, while others chose Isaac Newton or Albert Einstein. However, a very strong argument could be made that Charles Darwin is the most influential personality of the modern age. For it is Darwin more than anyone else who has changed the way humans view themselves, and he called into question the most cherished convictions of millions of people. Midway through the 20th century, college students were taught that Darwin, Karl Marx, and Sigmund Freud were the crucial figures of the age. Now only the reputation of Darwin is still intact and has, if anything, grown, while Marxism has been rejected in most parts of the world and Freud is more admired by poets than psychologists.[1]

CHARLES DARWIN (1809–1882)

Darwin's early life hardly suggested an iconoclastic personality. He was born in Shrewsbury, England, into comfortable circumstances the same day that Abraham Lincoln was born in a log cabin in Kentucky. Darwin's father was a successful physician with a substantial private fortune, while his mother was a member of the Wedgwood family, wealthy and famous for their pottery. His grandfather, Erasmus Darwin, was also a physician, but a libertine and free thinker, as well as an inferior poet. Erasmus's long, boring poem outlining his own theories of the emergence of species attracted relatively little attention. Darwin's mother was a genteel woman, a Unitarian noted for her piety, despite her father-in-law's definition of Unitarianism as "a featherbed for falling Christians."

Charles Darwin showed little early promise. When he attended boarding school at the age of 11, he did not in any way startle his tutors, who considered his academic potential ordinary. Even his father, who expected his sons to follow a profession despite the family affluence, feared he would amount to little. Left to himself, his father believed, young Darwin would choose the life of a hunting, shooting country gentleman, bringing no distinction to the family.

Without a clear vocational goal, Charles was sent by his father to the University of Edinburgh, where it was decided that he would study medicine, following the family custom. But Darwin soon proved himself unsuited to the practice of medicine; he could not endure the sight of blood or human suffering. Within a year, he had given up medical studies and instead busied himself collecting beetles. Declaring him unfit for all other professions, his family dispatched him to Cambridge University to study for the Anglican ministry. Though he was not especially religious, the life of a country parson would provide respectability, along with plenty of time for his nature explorations and collections of insect specimens. Important British landowners frequently gave their estate to the first son, sent the second one to the army, and found the third a place in the church.

As Charles Darwin pursued his studies, he read William Paley's *Evidences of Christianity* and found its arguments convincing. He agreed that, just as a good watch implied a skilled watchmaker, the universe showed forth the glory of its creator. Science was largely an amateur endeavor at this time, often pursued by clergymen. Young Darwin came under the influence of the Reverend John Henslow, who was an enthusiastic botanist in his spare time. But two other influences led Darwin to question received wisdom. Charles Lyell's *Principles of Geology* (published in 1830) revealed an ancient earth with defined geological epochs of enormous age. The other robust influence was the writing of Thomas Malthus, a clergyman who, in 1798, had published his *Essay on the Principle of Population*. Malthus had pointed out that the resources of the earth are limited and that food production could not keep pace with the growth of unrestrained plant, animal, and human populations. Natural disasters such as famine and disease, along with warfare, serve to hold population growth in check. Late marriage and other social conventions of the responsible classes provide some check on human population. Without constraints, living things would produce more offspring than the environment could ever sustain. As Darwin formulated his views and interpreted his observations, the writings of Lyell and Malthus were germinal.

Charles Darwin returned from university, perhaps at loose ends. Upon the recommendation of his former mentor, the Reverend Henslow, he was invited by Captain Robert FitzRoy to be his companion and resident naturalist on the voyage of the H.M.S. *Beagle*. Charles's father was adamantly opposed to the adventure, still fearing his son was becoming merely a dilettante, and it was only with the encouragement of a more tolerant uncle, Josiah Wedgwood, who would later become Charles's father-in-law, that the elder Darwin relented. The voyage, which was to last three years, left England in 1831. Its commission was to map the coastal waters of the southern part of South America. But there was also a religious intent of the voyage, because a native of Tierra del Fuego, who had been educated in England, was returning as a missionary to his own people. While on the ship, Darwin read his Bible and looked for churches to attend during shore excursions.

Captain FitzRoy was not always a genial companion. During one shore excursion, Darwin, always sensitive to suffering, was a horrified witness to a slave auction in Bahia, Brazil. An impassioned argument with FitzRoy over slavery followed. But Darwin was also disturbed by the primitive behavior of the tribes he encountered on other brief trips to the mainland. The inhabitants of Tierra del Fuego, for example, seemed to him more animal than human in their habits. Still, Darwin was overwhelmed by the splendor of the South American vistas. He collected fossils and kept a careful notebook of the natural wonders he observed, especially when he visited the Galapagos Islands. His observations of the distinctions between similar flora and fauna on the different islands of the chain—particularly the finches and the turtles—became important when, back home, he started formulating his theories of evolution.

Three years after Darwin's return from this decisive voyage, he married his first cousin, Emma Wedgwood. This was to be a long, happy marriage; the pair were devoted to one another and deeply loved the 10 children born of their union. Like all the Wedgwood women, Emma was religious, and Darwin's hesitancy to offend her sensibilities was a major reason he delayed in publishing his theories, so long that he almost lost their priority. Though she remained a loyal supporter of her husband throughout the subsequent turmoil his ideas generated, Emma retained her faith, along with her greatest fear, that she and Charles might be deprived of one another in an afterlife.

In the years following his voyage, the Darwins lived in the hamlet of Down, where he refined his theories and collected ever more evidence to support them. As he contemplated the writings of Lyell and Malthus, he also closely observed the selective breeding of farm animals. Despite fragile health, he worked industriously. To the end of his life, he modestly claimed that his only talent was "an unusual power of noticing things which easily escape attention, and of observing them carefully." He might also have acknowledged his clear, logical, and sometimes even eloquent writing style. Still, he hesitated to publish his theories, no doubt fearing the uproar that eventually ensued, though he carefully prepared a record of some two hundred pages, which he instructed his family to publish in case of his death. Suffering bouts of illness, he doubted that he would live long enough to complete his work.

Everything changed in 1858, when Darwin received a letter and scientific paper from Alfred Russel Wallace, a biologist he knew only casually. He discovered from that communication that Wallace had independently arrived at many of the same conclusions as he had reached. Darwin, who had already revealed his theories to a few close associates, was amazed at the coincidence. Not only were Wallace's views on natural selection and the struggle for existence almost identical to his own, but Wallace had even used phrases that Darwin had placed at the head of the chapters of his own manuscript. Not wanting to be deprived of recognition for his originality by his fellow scientist, Darwin consulted friends, including Charles Lyell and Joseph Hooker, to whom he had earlier confided his theories. Their proposed solution was quickly accepted by Darwin and by Wallace, who had long been Darwin's admirer. It was arranged that papers by Darwin and Wallace would be read jointly before the Linnaean Society, so that both men would receive credit for the theory.

After this, and with Wallace's blessing, Darwin was no longer hesitant to publish his writings. In *On the Origin of Species* of 1859, he was still careful not to tackle head-on the thorny question of human origins, and he even expressed a conventionally pious sentiment at the conclusion of the book. But the implications of his work for human origins were clear to his careful readers, and in 1871 he published his second most important book, *The Descent of Man and Selection in Relation to Sex*. It was this later book that presented the part of his theory that was truly original and critical—his mechanism for evolution, the role of sexual selection in the struggle for survival.

Not all of Darwin's views would hold up to later scrutiny. For example, he accepted Lamarckian genetics, which taught that acquired characteristics could be passed on to offspring. Despite earnest efforts of researchers in laboratories with fruit flies and toads, Lamarckianism later proved impossible to verify. The long necks of giraffes, a favorite Lamarckian example, it turned out, were not acquired because their ancestors had snipped food from tall trees. There were also gaps in Darwin's system that others would be free to explore. Yet the basic outlines of his theory would be accepted with remarkable rapidity by scientists.

Other books followed. In 1872, Darwin published *The Expression of the Emotions in Men and Animals,* further suggesting that humans differ from lower animals only in degree and not in basic nature. In 1881, he published his last important work, *The Formation of Vegetable Mould through the Action of Worms,* which appears to have been his own favorite book, because it took him back to the collecting and classifying he had most enjoyed in youth.

Darwin never delighted in challenging the cherished beliefs of others. But by the time his books were receiving wide attention, his religious views, quite orthodox during his university years, were no longer in any way conventional. He had long before left Anglicanism and finally could no longer accept even the Unitarianism of members of his family. Most of his biographers feel that his final flight into agnosticism resulted less from his research and his theorizing than from the death in childhood of Annie, his beloved daughter. Increasingly, Darwin perceived the cruelty of nature, what Alfred, Lord Tennyson referred to as "nature red in tooth and claw." And as his health further deteriorated, perhaps as a partial result of the anxiety he experienced, Darwin came to acknowledge a cruel edge of intolerance in Christianity itself, though he admired the high ethics of the faith. With him, as it usually is with lesser folk, the retreat from religion was more emotional than intellectual. Any hope for the existence of a benevolent Creator vanished, as members of his own family lost out in the struggle for existence.

Darwin still donated to religious causes, even evangelical ones. This was in part a mark of his respect for the religious sentiments of his dear Emma. After his death, as was not uncommon when famous people died outside the church, many deathbed stories of conversion circulated. All of them were almost certainly without foundation. If they had been true, Emma would certainly have acknowledged them. Nevertheless, a celebrated evangelist of the day, a Lady Hope who toured England and North America, claimed to have visited the great man in his last moments. As the years passed, her story took on added detail, and she appears to have convinced herself that she had brought Darwin back to Christianity at the last moment. Though she may indeed have visited him in his final years, his family never verified her conversion accounts.

By the time Darwin died, the upheavals he had wrought no longer touched him so personally, and his reputation was already so solid that he was honored with a burial among kings and other greats in Westminster Abbey, not far from the tomb of Newton. In 1885, his statue by Sir J. E. Boehm was unveiled in the National History Museum in London in a ceremony attended by dignitaries and divines, including the Archbishop of Canterbury and the Prince of Wales.

ALFRED RUSSEL WALLACE (1823–1913)

Darwin's colleague and co-originator of evolutionary theory was also an Englishman, one whose background and life circumstances were radically different from his own. While Darwin was born to wealth and privilege and came from a family already distinguished, Alfred Russel Wallace was born in poverty and obscurity. He did not have Darwin's educational opportunities nor the same leisure to pursue gentlemanly research. Wallace worked for a time with his brother as an apprentice surveyor, but he endured periods of dreary unemployment. He was largely self-taught but was intellectually impressive enough to be employed for a time as a master at the Collegiate School in Leicester, where he taught drawing, mapmaking, and surveying. During that time, he continued his program of self-study and became acquainted with the work of Malthus on population, which made a deep impression on him, just as it had upon Darwin. Wallace also made friends with a young entomologist, Henry Bates, who was already establishing himself as an authority on beetles. This encouraged Wallace to begin his own collection of insects. He was soon able to leave his teaching post to work as a civil engineer on a railway system, allowing more opportunity to add to his collections.

Wallace, inspired by the writing of naturalists who worked in foreign lands, longed to travel. He accepted an offer to visit Brazil, where he explored parts of the rain forest, with its rich assortment of flora and fauna. Unfortunately, the valuable specimens he collected there were destroyed in a fire on the ship as he returned to England. In order to expand his opportunities for research, he then accepted work that took him to the East Indies, to the lands now known as Malaysia and Indonesia. His original work in zoological demarcation led to the identification of what has become known as "the Wallace line." In 1869, he published an account of his travels entitled *The Malay Archipelago,* which he dedicated to Charles Darwin, whom he had met only once but greatly admired.

In 1866, Wallace married Annie Mitten, the daughter of a recognized authority on mosses. The Wallaces had three children but lost one in childhood, not unusual for the time but no less heartbreaking. Because the financial affairs of the family were always precarious, Wallace set up a small business selling specimens that he had collected during his travels. Though he had admirers in the scientific community, he was never able to secure a permanent position in areas of his scientific expertise. Sadly, he remained the gifted amateur, eking out a living for his family by grading government examinations and editing the writings of people such as Lyell and Darwin. Aware of Wallace's financial difficulties, Darwin befriended him several times and used his influence with the government to secure a pension for Wallace, which was finally awarded in 1881.

Wallace had developed his own theory of the transmutation of species during his travels. Working in the Amazon, he had observed how geographical barriers, such as mountain ranges and rivers, separated closely related species. He studied geographical and geological distributions of species, as he moved from place to place. In his autobiography he later described how, lying in bed with a fever, he had contemplated Malthus's writings on the checks and balances to human population and reached his own conclusions about natural selection.

Though Wallace was known for his originality, creativity, and ability to generate new insights, his tentative theories tended to be dismissed by leading scientists such as Georges Cuvier, Adam Sedgwick, and Charles Lyell. He had nevertheless kept an open channel of communication with Darwin, whom he always regarded as friend rather than rival.

Because Darwin had a larger scientific reputation and, in class-ridden British society, higher social status, their mutual discoveries and ideas received more respectful attention when presented by Darwin rather than by Wallace alone. Both men subsequently were recognized as co-discoverers of the mechanism of evolution, though Darwin's more substantial later publications solidified their work and supported their theories more fully. The two men remained friends and respected colleagues for life.

It is not surprising, given the public's love of conspiracies, that in the 1980s, there was an attempt to exalt Wallace's reputation in books by Arnold Brackman and John Langdon Brooks. These writings maligned Darwin's character, suggesting that Wallace had been cheated out of his due credit by a conspiracy of elite establishment scientists. Darwin was even accused of stealing his central concept of evolution from Wallace. Brackman and Brooks, however, presented no convincing evidence, and further investigation by other historians of science has discredited their thesis.

Wallace did not follow Darwin's path into agnosticism. In his later years, he became interested in phrenology, hypnosis, and spiritualism. He attended séances and believed some of them were genuine. His defense of a few spiritualist mediums against charges of fraud damaged his own scientific reputation. Thomas Henry Huxley, always a partisan of Darwin, pronounced Wallace a crank. However, Wallace's contribution to evolutionary theory, even his first use of the phrase "survival of the fittest," could not be seriously challenged.

EARLY THEOLOGICAL RESPONSES TO EVOLUTION

Theological responses to Darwin's books and ideas were swift and heavy. Not only was he defying established science—let the specialists deal with that—but he appeared to be attacking the very foundations of religious faith. He was also challenging the secular humanism of the West, inherited from the classic Greeks and Romans—the belief that "man is the measure of all things."

Of the early critics of Darwin, Samuel Wilberforce, Bishop of Oxford, was the best known. He came from distinguished British lineage and was renowned for his learning and oratorical skills. In the influential *Quarterly Review* (July 1860), he argued that Darwin was attempting to limit the glory and power of God in creation, that his views were totally incompatible with the revealed word of God, and that he challenged what religion had always taught about the special relationship of humans to their Creator. Wilberforce believed that the fall of Adam was responsible for the many strange forms found in nature and the vicious struggle for existence that all living things endured. As an Anglican divine, Wilberforce was a spokesman for the Established English Church, but Roman Catholic prelates in England were equally severe in their denunciations.

Across the channel in France, intellectuals prided themselves on their belief that they were the most rational of people, and they did not especially enjoy honoring Englishmen. The initial revulsion to Darwinism was, perhaps understandably, stronger here than in Britain. Several religious dignitaries denounced Darwin. Fabré d'Envieu stated that any doctrine other than the fixity and persistence of species contradicted Holy Scripture and was absolutely anathema. Abbé Desorges, an eminent professor of theology, pronounced Darwin a "gloomy pedant," while Monseigneur Segur attacked the teachings of Darwin and his followers with heated words: "These infamous doctrines," he wrote, "have for their only support the most abject passions. Their father is pride, their mother impurity, their offspring revolutions. They come from hell and return thither, taking with them the gross creatures who blush to proclaim and accept them."[2]

In Germany, Protestant and Roman Catholic theologians seemed to compete with one another in the ferocity of their denunciations. Darwin's theory was called "a caricature of Creation." One theologian said Darwin had "turned the Creator out of doors," while another protested that "Every idea of the Holy Scriptures, from the first to the last page, stands in diametrical opposition to the Darwinian theory."[3] In German-speaking Switzerland, there was a call for no less than a full intellectual "crusade" against Darwinism.

Reaction in the New World was initially more restrained. Still, the theologians who kept abreast of scientific developments were not ready to fully capitulate. Darwin was widely denounced as an infidel, "sophistical and illogical," and an obvious threat to morals and civil order. Furthermore, Americans quickly seized upon evolutionary ideas as a threat to sound education and good behavior. Critics questioned how children could be prevented from behaving like monkeys if taught they were related to them. The fact that Darwin never taught or believed that humans were descendants of monkeys was irrelevant, so distorted were his writings in the popular mind. In Australia, Dr. Charles Perry, Lord Bishop of Melbourne, bitterly identified both Darwin and his "bulldog," Thomas Henry Huxley, as persons consciously attempting to undermine the Christian religion and its Holy Scripture.[4]

Darwin's second important publication, *The Descent of Man,* was, of course, more troubling than his first, which had made some concessions to religious sensibilities. The uproar attending this publication was immediate. From Ireland came the suggestion that the book was so outlandish that it must certainly have been intended as satire, an extended ironic essay in the spirit of Jonathan Swift's "Modest Proposal" or Erasmus's *Praise of Folly.* In England, Prime Minister William Gladstone, known for his evangelical piety, was equally eloquent and passionate in his tirades.

Rome's highest authority, Pope Pius IX, spoke of Darwinism as an "aberration," repugnant to history, to the tradition of all peoples, observable facts, and reason itself. Darwinism was even a perversion of science. It was, according to the pope, "so fantastic a fairy tale, as it were, that it would have received little attention had the modern world not been already so far gone in materialism and depravity."[5]

Some religious thinkers were repelled by the doctrine of survival of the fittest, which they saw as an especially strong challenge to the concept of a just and merciful God. Could such a Creator as revealed in the Christian tradition set in motion a plan that would reward ruthless aggressiveness? Did this not contradict the teachings of Jesus that the meek would inherit the earth? Still, there were others who found some support for Darwin's grim vision in holy tradition. Had not the patristic theologians taught that all creation suffered the effects of original sin? Some had even suggested that the earth's tilt occurred at the moment of Adam's fall, so that now harsh seasons punished much of the earth and the people who inhabited those regions. Had not all of nature revolted against God? Still, it would take time and deep reflection before religious scientists and theologians would be able to reconcile evolution with their faith and before even a pope would be able to affirm evolution.

THE DARWINIAN PRECURSORS

Darwin's theory had an elegant simplicity; it was not too difficult to understand, and the similarity of species along with the results of domestic animal breeding appeared to lend support. Yet it was so revolutionary in its ramifications that it shook both religious and scientific establishments and caused distress in the general public. For it questioned the existence of a higher purpose in human life, in all living things, in the universe itself. Before Darwin, most people believed that life was purposeful, even if individual lives often seemed shrouded in darkness. After Darwin, as Richard Dawkins would later recognize, atheism became intellectually respectable. Today many educators assert that the ultimate questions of the universe are metaphysical and not a matter for scientific verification. However, the clear implication of much of modern science—Darwinism especially—seems to be that there is no discernable pattern or orderly progression in the universe.

Before modern science gained its present prestige, most of the Western world based its view of history and civilization on revelations believed to be divine. The Bible dealt with human history, from its beginnings to its predicted end. And many people were content to accept the age of the earth, as Bishop James Ussher in the 17th century had calculated it from Hebrew genealogies, to be only a few thousand years old. But there had been some foreshadowings of Darwin.

In the 17th century, a French Calvinist, Isaac de la Peyrere, had put forth an interesting theory derived from both the Bible and his own fertile imagination. He concluded that before Adam, a race of Gentiles had already existed. This theory was even revived in England about the same time as Darwin's *On the Origin of Species.* In more recent years, a few Jewish and Christian scholars, who find it necessary to reconcile evolution with a more or less literal reading of scripture, have postulated a pre-Adamic race that lacked the human soul breathed into Adam, the first complete human.

Progress, progression, and hierarchy were important in pre-Darwinian thinking. The Great Chain of Being, a favorite image of philosophers and poets, was a blend of pagan Greek with later Christian thinking. God, representing perfection, existed at the top of this chain, or ladder. Angelic beings, in their own hierarchies, and humans, with their social ranks, were lower. Renaissance thinkers even found places for animal species on different rungs, with the lion, king of the beasts, ahead of all others.

But the Enlightenment proposed other ideas. Species had not necessarily been static since Creation. Benoit de Maillet in France published a work suggesting that life had developed from simple to complex forms. He was a diplomat who had broadened his outlook through service in Asia and Africa. According to his calculation, the earth was some two billion years old, with humankind making its appearance about 500,000 years before his own time. When he published these ideas in *Telliamed* (1748), the public found him too radical, and his work was roundly rejected.

In the middle of the 18th century, the Comte de Buffon theorized that the earth and the other planets had expanded from fragments chipped off the sun during a gigantic cosmic collision. He believed the earth had gradually cooled so that life had appeared, first in the sea, after about 33,000 years. Animal and plant life came about 60,000 years later. Humans, he calculated, must have emerged after 70,000 years. Buffon spoke of ancestral forms of life changing in response to environmental conditions, thus foreshadowing Darwin. He was also one of the first to speak of species as clearly defined breeding groups.

About the same time, an astronomer, Pierre Laplace, suggested that everything came from a nebular cloud. Laplace was a Deist who believed God had created the cosmos so skillfully that he did not need to constantly manage it. Thus, the 18th-century intellectuals had already introduced theories that strongly differed from the biblical account of creation, though millions of religious people were still untouched by them.

Biological science developed and gradually became professionalized in the 19th century. In the German lands, both Carl Friedrich Kiel Meyer and Friedrich Schelling speculated on the possibility of modest evolution, though it is unclear to what degree they fully accepted the idea. In England, Charles Darwin's own grandfather, Erasmus Darwin, a Deist, outlined his primitive evolutionary ideas in doggerel. England, which had very good poets, paid little attention to his writings.

In France, Jean-Baptiste Lamarck published a very important book in 1809 entitled *Zoological Philosophy*. He detailed an evolutionary process within a theistic framework. God in his wisdom, according to Lamarck, determined in the beginning how life would appear and progress. An *élan vital* directed organisms toward ever more complex development. Lamarck became especially well known for his theories of heredity, which were widely accepted for a time and would die a slow death.

In 1844, an anonymous work, *Vestiges of the Natural History of Creation,* appeared in Britain and attracted much discussion among intellectuals. The book took many of the ideas of Laplace and painted a grand tapestry of life constantly developing over the eons. These ideas were attractive to many English people because *Vestiges* also appeared to show a progression in human civilization, flattering to Europeans and in harmony with the jingoistic thinking of the time. All these works helped pave the way for Darwin.

Evolutionary concepts were received by some only cautiously, if at all, for several reasons. Religion was only the dominant one. For some intellectuals, Deism had already made of God an absentee landlord, and evolution further separated humanity from him. Lamarck's influential system had rejected the idea of the extinction of species. Though fossil remains were beginning to attract attention, their study was not yet advanced enough to clearly detect species that no longer inhabited the earth. Because earlier species were often quite similar in bodily structure to existing ones, there was little observable evidence that animals had changed over time. Species might vary, as in farm breeding, but within well known limits. Evolution was further associated in the popular mind with revolution, and England feared the sort of turmoil that French society had endured. Theories that removed God from the scene were not popular, because they were too reminiscent of the revolutionary thinking of French philosophers.

Darwin generously credited his predecessors, tried to correct their limitations, and openly acknowledged the gaps in his own theory. He had carefully accumulated his facts, anticipating objections sure to be made and questions that would be asked. Where data were lacking, he was nevertheless confident that later research would confirm his findings. Among the difficulties he identified was the absence of missing links because of the slim evidence of the fossil record. Yet his theory had obvious strengths. Natural selection helped explain the geographical distribution of species. It was also consistent with what was known from other scientific research, providing explanations for much that had before seemed mystifying.

DEVELOPMENTS FROM DARWINISM

Several important scientists quickly built on Darwin's findings. In Germany, Karl Vogt promoted Darwinism is his university lectures, and Darwin's leading disciple, Thomas Henry Huxley, published *Man's Place in Nature* (1863), which did much to spread an understanding of human evolution. Meanwhile, paleontology was emerging as a professional science. Fossils were being discovered in many parts of the world. In the 1890s, Java Man was much discussed as a possible missing link between humans and their animal ancestors, but it was later dismissed as such. The writings of the German Darwinist Ernst Haeckel encouraged the search and discovery of many important fossils. One mystery that was frequently debated, and has yet to be definitively solved, had to do with the geography of human origins. Did humans originate in a single spot (the "out of Africa" hypothesis), or did they emerge more or less simultaneously in several parts of the world? Not until the end of the 20th century would most paleontologists come to favor Africa as the original homeland of the human race.

Several German intellectuals were particularly enthusiastic about possible social implications of Darwin's theories. Even after acknowledging the force of Darwin's arguments, some were still not ready to demote humans to the status of merely a higher animal. Ludwig Buchner in 1855 published *Force and Matter,* in which he affirmed that evolution had indeed occurred but that humans were the highest products of the process and were, therefore, immune to any further evolutionary development. He believed that the social institutions humans had developed would distance them from the further struggle for existence. Like numerous other 19th-century thinkers, he was ready to incorporate some of Darwin's discoveries into a progressive credo of his own, which Darwin himself could never have accepted.

THE NEO-DARWINIAN SYNTHESIS

Because the science of genetics had not yet been structured, developed, or even named, Darwin had difficulty reconciling his theory to hereditary change. He put forth a concept of pangenesis, attempting to explain heredity by means of material particles that carried information from one generation to another. This theory permitted the inheritance of some acquired characteristics, but it left too much unexplained and could not be verified.

The majority of people who accepted Darwinian ideas by the end of the 19th century still believed in some sort of purposeful direction to evolution, a goal at the end of the process. Darwin's outline of natural selection had fallen out of favor. Herbert Spencer, a very influential thinker who coined the phrase "struggle for existence," found natural selection inadequate. George Bernard Shaw, a famous playwright with an opinion on everything, still relied on Lamarckian theories of inheritance. Liberal theologians who had decided to accept some evolutionary theory still tended to discount natural selection. Orthogenesis, which gained some popularity in the United States, taught that evolution followed a preset course, thus making it more congenial to Christian teachings.

It was the work of the Austrian monk Gregor Mendel that most historians of science credit with rescuing Darwinism from a serious decline in the early 20th century. Mendel's research with pea plants in his monastery garden led to his identification of the basic laws of heredity. His work was published in a scientific journal and then largely forgotten. Although Mendel had sent some of his papers to Darwin, hoping his research would help Darwin verify his own theories, it is not clear that the pages were even opened. It was left to later Darwinians to rediscover Mendel's work, leading to a more persuasive neo-Darwinian synthesis that would emerge in the 20th century.

When Mendel's essential work was finally rediscovered in the early 20th century, Hugo de Vries coined the term *mutation* to describe large variations that appeared and could be inherited. William Bateson, who questioned the power of selection alone to produce new species, was a dominant figure in the developing science of heredity, which he named genetics. By 1920, it was becoming clear that Mendel's work and the subsequent research it generated were crucial to a rescue of Darwinism from obscurity.

An intriguing and tragic figure in the history of genetics is the Austrian research biologist Paul Kammerer. A genius with many talents, Kammerer was a musician of professional caliber. He was a dashing figure, loved by many women. Sacrificing a career in music, he decided to devote himself to scientific research, hoping to confirm Lamarckian genetics—already much on the defensive—by proving that acquired characteristics could be inherited. His experiments with the midwife toad attracted international attention. This rather peculiar creature has an unusual means of reproduction; while most toads mate in water, the midwife mates on land. After fertilizing the eggs of the female, the male winds them into long gel-like strands around his hind legs and carries them until they are hatched. Unlike toads that mate in water, the midwife does not have nuptial pads, blackish-colored palm swellings fitted with small protruding spines, which enable the male to clasp the female in water. Kammerer believed that by inducing midwife toads to mate in water, contrary to their usual habit, he could, after a few generations, demonstrate that they acquired nuptial pads, which were then passed on to their own young. He exhibited some interesting specimens with blackened padlike growths, which he felt verified Lamarckian heredity principles. Agreeing to show his specimens in England, he was confronted there by the leading Mendelian, William Bateson, a ferocious opponent. The exhibits were exposed as fake; the nuptial pads had been colored with black ink. Kammerer was pronounced a charlatan, though some of his biographers believe he did not intentionally falsify his research but was the victim of someone in his laboratory who sought to discredit him. Though Kammerer was offered an important position at Moscow University, which still accepted Lamarckian heredity, he committed suicide in 1926, just before his anticipated departure for Russia. Mystery still surrounds his career and the end of his life.[6]

Stalin's Soviet Union remains a cautionary tale of the harm that results when the state determines how scientific research will be conducted. Before 1929, Soviet scientists were productive and highly regarded by their colleagues in the West. Stalin, however, formulated all policies, scientific and otherwise, as his power consolidated. Every intellectual discipline had to conform to his interpretation of Marxist-Leninist ideology. Trofim Lysenko, a man of peasant origins and thin academic credentials, became his favored scientist. Lysenko, a largely self-trained agronomist, experimented with the germination of winter wheat, a very important crop in Russia. Denouncing the new science of genetics as a "bourgeois product of the decadent West" and relying on Lamarckian principles of heredity, Lysenko developed a process called vernalization. Believing he could freeze spring wheat and give it the desired characteristics of winter wheat, Lysenko claimed these acquired features could then be passed on to subsequent generations of wheat, providing food for millions of Soviet citizens. Before the full disaster of his procedure could be revealed, Lysenko, presented as a peasant genius, rose to prominence in Stalin's government, becoming a high Soviet official.

There were several reasons the Soviet government preferred Lamarckian theories to the more solidly established ones that had gained acceptance in the West. First, Western genetics appeared incompatible with any form of progressive evolution. It could not be reconciled with the Marxist faith in a coming egalitarian society. Lamarckian theories, on the other hand, suggested a purposeful evolutionary process, highly compatible with the goal-directed theories of communism, in which the emergence of an ideal society in the future was predicted.

By the 1930s, most issues of heredity had been clarified and a clear neo-Darwinian synthesis was widely accepted by Western science. Population geneticists and naturalists revived a flagging Darwinism. Theodosius Dobzhansky, a Russian working outside the restrictive Soviet Union, explained how geography affected the genetic complexion of a population. His work in turn influenced Ernst Mayr, who studied geographical isolation and its role in the emergence of new species. Paleontology further revealed many new findings that contributed to this synthesis. George Gaylord Simpson combined the work of paleontologists with genetics to explain macroevolutionary factors. The new synthesis excluded Lamarckian theory, which should have vanished long before through clear observation. (Had not the Jews practiced circumcision for millennia without bodily changes showing up in offspring?) Molecular biology and the later findings in DNA research helped complete the picture. Now it became extremely difficult for scientifically minded people to challenge the reality of evolution.

NOTES

1. For the material in this chapter I am deeply indebted to the extraordinary lectures of Professor Frederick Gregory of the University of Florida, for The Teaching

Company, 2008. These lectures, titled "The Darwinian Revolution" are available on CD and DVD from The Teaching Company, Chantilly, Virginia.

2. Andrew Dickson White, "The Final Effect of Theology," in *Darwin,* ed. Philip Appleman (New York: W.W. Norton, 1970), 425–426.

3. Ibid., 426.

4. Ibid., 425.

5. Ibid., 427.

6. For a full account of this minor but intriguing footnote to evolutionary history, see Arthur Koestler, *The Case for the Midwife Toad* (New York: Random House, 1972).

3

Social Darwinism and Eugenics

Social apprehensions added to religious reservations as Darwin's ideas spread throughout Europe and the United States. Although never advocated by Darwin himself, programs that became known as social Darwinism and eugenics appeared to many the natural extensions of evolutionary theories. Embraced by social theorists who felt the human race could be improved, even saved, by applied Darwinism, these programs eventually became some of the strongest impediments to the popular acceptance of evolution.

Darwin, a gentle man of science, should not be indicted for the distorted applications that others would soon make of his theories. Though he believed, as did most Englishmen of his time and social class, in the natural superiority of the Anglo-Saxon peoples, he cannot be held responsible when "survival of the fittest" was used to justify imperialism, colonialism, militarism, and violent racism.

Laissez-faire capitalism in its raw form attempted to justify its practices by appealing to a "natural order" in which people of lesser gifts could be validly exploited as poorly paid labor to enrich those of superior endowment. The best economic system was thus identified as the one that operated according to competitive market principles, without government interference. Influences on the market itself were deemed sufficient to keep goods and services flowing, while limitations imposed by a government were destructive. In the United States, "robber barons" accumulated huge fortunes, while their workers remained mired in poverty. The Industrial Revolution in England had caused vast social disruptions, with peasants flooding into crowded cities infested with disease and crime and with the entrepreneurial class gaining enormous profits on the backs of people, including children, who worked from morning to night for a pittance. Although the heartlessness of the system was evident to citizens of conscience, the poorer classes were still widely regarded as inferior, childlike beings, wallowing in drunkenness and debauchery. Attempts to put restraints on capitalism and alleviate the state of the poor were, consequently, often met with objection from those who claimed the "bleeding hearts" were interfering with the natural order. Just as Europe experienced the disruptions of the Industrial Revolution, the United States during the early years of the 20th century struggled to absorb waves of immigration from countries whose customs were alien to most Americans. There were fears that these new arrivals, deemed naturally inferior, could not easily assimilate.

Marxist theorists used the same class-based thinking to point to different ends, as they chose from Darwinism what they found useful to the economic system they proposed. Unlike Darwin himself, they prophesied a social progression in the struggle for existence. They believed that class conflict led inevitably to revolution, and a classless society would eventually emerge in which workers would prevail. In developing communist ideology, Friedrich Engels and Karl Marx were much taken with several Darwinian ideas, which they interpreted according to their own philosophy. To them, all history revealed this class struggle, with successive upheavals from time to time destroying the old orders. The slavery of classical ages had given way to the serfdom of medieval feudalism, which in turn had led to the supremacy of the bourgeois capitalists. This progression would continue through a worker's revolt to the triumph of a classless society. The old stratifications of society would give way to the ideal society they envisioned, where work would be its own reward, and everyone would take what was needed and contribute what was required. Their sympathies were with the exploited poor. Their theory, as true Darwinians would have quickly observed, was fallacious in that it did not take into account human nature and the motivating forces that bring about creativity and productivity. Yet these early socialist philosophers were so enthused with their theories of class struggle and its seeming support from evolutionists that Marx wanted to dedicate *Das Kapital* to Darwin, who politely declined the honor.

Herbert Spencer, whose ideas were influential in the latter half of the 19th century, was convinced by Darwin's work that government attempts to mitigate this social struggle would be a mistake. Walter Bagehot accepted social Darwinism in England, and William Graham Sumner promoted it in the United States. From his professorial post at Yale University, Sumner taught that social inequalities resulted from the different abilities and intellectual powers that people were born with. The struggle of life itself would eventually eliminate the ill equipped, while those with the best racial heredity, health, sound morality, and cultural strength would survive. His ethic entailed hard work, thrift, and sobriety, and he described inferior classes as lazy, drunken, and financially irresponsible. With these opinions firmly held, he rejected government efforts to alleviate the plight of the poor, whom he believed could not be persuaded to adopt the standards he upheld, subsequently known as "middle-class values."

The social Darwinists who emerged in the late 19th century believed also that certain "inferior races"would be subject to natural selection, and would be weeded out in the continuing struggle. Today it is often forgotten how prevalent racists doctrines were at this time among Europeans and much of the rest of humanity. Nations, too, were caught in this battle for survival, it was believed, and those countries that did survive would have proven their superiority. Since the early social Darwinists were white, primarily central or northern Europeans, they concluded that theirs were the advanced civilizations that deserved to rule over those they deemed less fit—primarily Asians and Africans. Thus came the justification for colonial exploitation and imperialism. The British spoke of "the white man's burden," while Americans, though somewhat less colonially adventurous, proclaimed a "manifest destiny." Germans, of course, at this time regarded themselves as the most civilized among humankind. The French disagreed, promoting their own culture as the most refined. In the United States, despite its Constitution and multiplicity of peoples, social Darwinism still had a following. President Theodore Roosevelt gave it some credence and seems to have believed that white Americans would eventually submerge native indigenous peoples, while powerful nations would eventually subjugate more vulnerable ones.

The chief reason social Darwinism fell out of favor in the United States and Western Europe was its success and popularity in Germany, soon to become the enemy of England and the United States through much of the 20th century. The evolutionary biologist Ernst Haeckel, who in recent years has been found to have fabricated some of his highly publicized research, taught that advanced species had passed from primitivism through several higher states, suggesting another non-Darwinian elevating direction to evolution. Always a more a theoretical anthropologist than an experimental scientist, Haeckel set up a taxonomy of races, with Germans at the top. Races, he taught, differed in abilities and in their propensity to reason. The lower races, unlike civilized central and northern Europeans, were closer to the jungle and the animal state; therefore, not all human lives were of equal value. Some races and nationalities had the right to dominate lesser ones. Prefiguring Adolf Hitler, Haeckel placed Jews among the inferior peoples, despite the fact that Jews had not only been historical achievers but even within Haeckel's Germany had made contributions far outweighing their numbers. Possibly in a misguided attempt to influence Christians, Haeckel taught that the founder of the Christian religion, Jesus, had not really been a Jew. Haeckel was actually borrowing an ancient Talmudic calumny and claimed, with absolutely no historical evidence, that Jesus was the son of a Roman soldier. The dire effect of his thought would not be fully realized until his discredited theories were resurrected during the Nazi era.

The disciples of Haeckel were fond of examining the skulls of Africans and of women of any race, attempting to find proof that these brains were inferior and juvenile in form. So, not only were they able to justify racism and colonialism, but they also thought they were providing a clear rationale for the subservience of women. Unscientific as these ideas have proven to be, some of them have lingered on and have not entirely disappeared in the first decade of the 21st century.

Perhaps the most heinous misuse of Darwinism was in justifying war. If nature indeed, in Tennyson's words, operated "red in tooth and claw," and if this was an essential pruning of the earth for the domination of the fittest, what was wrong with war? In warfare, too, it was reasoned, natural selection would operate beneficially. With military conquest justified, a number of countries in Europe—Germany, France, Belgium, Portugal, England—set about forcibly colonizing populations of Africa and Asia, exploiting their resources. It came as a surprise to some of these Europeans to discover, in places like China and India, civilizations far more ancient and rich than their own. Eventually these ideas would turn on Europeans themselves, as Germans further justified their conquest of their geographical neighbors with these same bogus Darwinian principles.

Another reason Darwinism was so widely rejected in the United States was its association in the popular mind with eugenics, a pseudo-science popular in the early 20th century, given its name by Sir Francis Galton, a cousin of Charles Darwin. According to evolutionary theory, those who win in the struggle for existence are not necessarily the brightest nor the physically strongest. They are the ones who have reproductive success. And these prolific individuals, it turns out, are not always the most socially desirable. Surely humans with their reasoning abilities could take control of evolution and create a superior population, the eugenicists reasoned. It was widely noted that more gifted people usually do not reproduce as plentifully as those who are often judged unfit by most social criteria. The prognosis for society seemed dim, with the less able and intelligent overrunning the earth. Eugenics was a response to this "crisis"; though genocide was not advocated, eugenicists sought to improve human population through a more careful selection of the people who married and reproduced. If choice animals could be selectively bred, should not at least equal attention be given to the breeding of a superior human race?

Galton believed that those with high intelligence, talents, ethical upright-ness, and physical strength should be encouraged to reproduce in greater numbers. He accumulated a large amount of data that he believed sub-stantiated his theories and gained a small but influential following. Some of his disciples attempted to put his theories into practice in both Britain and the United States. Even the Protestant Episcopal Church at one time attempted to aid the program, and on at least one occasion it was sanctioned by the United States Supreme Court. Although eugenics champions had little success in convincing "the right people" to marry one another, the movement did make its gains, or so it was thought, in the widespread sterilization of persons believed to carry harmful genes. Eugenicists made appeals to history, claiming ample precedent for its program. The ancient Spartans had devel-oped a society of the physically robust by carefully selecting those who could live and those who would die. Yet while the Spartans proved sturdy in combat and knew how to endure austere living conditions, they never achieved the cultural glories of the somewhat more lenient Athenians.

Interest in eugenics spread as the heredity laws of Gregor Mendel became widely known in the early part of the 20th century. Elaborate studies were published of deficient families such as the Jukes and the Kallikaks. Hereditary genius was traced in the Bach, Mendelssohn, and even the Darwin families. A center for research in human evolution was established in the United States, at Cold Spring Harbor, New York, presided over by Charles Davenport, a biologist. Davenport was convinced that his program, if carefully followed, could eliminate alcoholism, feeble-mindedness, prostitution, and poverty within a few generations. He believed human pairings should be arranged by pedigree, much in the manner of horse breeding, in order to produce outstanding offspring.

During the first three decades of the 20th century, 32 states passed laws requiring sterilization of people determined, by questionable criteria, to be undesirable. The mentally ill, those with serious physical handicaps, and people who had been convicted of major crimes were considered degenerate, good candidates for sterilization. Several states also maintained laws against interracial marriage, which was thought to weaken the stock. These laws were very difficult to enforce, and common law marriages were common in many places, especially the southern states, where the races had long intermingled, even if unofficially. Still, the eugenicists preached their gospel with special fervor.

Among the prominent Americans who gave some support to eugenic pro-grams were—in addition to Theodore Roosevelt—Alexander Graham Bell and Calvin Coolidge. Margaret Sanger, a pioneer in the population control movement, also believed in encouraging parents deemed genetically privileged to have more children, while less able ones should have few or none. Not only did Sanger value intelligence, but she weighed athletic prowess, radiant health, and culturally conditioned notions of physical beauty. Since she, like most of the eugenics theorists, was of northern European stock, she believed this type was the proper model.

The ideas of Thomas Malthus were still influential. Malthus had written of the natural limits to resources and the struggle for sustenance by populations greater than the earth could support. The unbridled fecundity of humans as well as other living beings was painfully held in check by predators, disease, natural disasters, and wars. In this competition, winners would eliminate losers. Yet if governments could encourage humans to limit their offspring, much of this suffering might be averted. Alfred Russel Wallace, though no eugenicist, had mitigated his own theories of survival of the fittest with spiritual concerns. And he, too, believed that humans should use their intellect to direct the evolutionary process and create a more humane society.

A number of sociologists and psychologists lamented what they referred to as "race suicide," observing that people of achievement continued to delay marriage and had fewer offspring than folk of less status and ability. Eugenic societies lobbied for appropriate policies of mate selection and held "better baby" contests. Some countries even tried to provide incentives for professional people to have more children. These policies, not surprisingly, did not win mass acceptance in democratic countries. The United States, despite its fear of foreign elements, had, after all, a log cabin tradition and a Horatio Alger myth that the poor but virtuous and hard working could rise to the top. To its credit, the Roman Catholic Church, less influential in the United States than it would later become, opposed these eugenic policies from the start.

Still, every American state and most Western nations were in some way influenced by eugenic ideas, sometimes passing laws sterilizing "undesirables" or directing policies that sexually segregated people who were labeled inferior. The case of Carrie Buck is especially notorious and a black mark on U.S. history. Buck's plight reached the Supreme Court in 1927, in *Buck v. Bell.* The state of Virginia had declared Carrie Buck, a 17-year-old woman, "feeble-minded" and had forcibly sterilized her in the 1920s. She had been told that her surgery was for appendicitis. When details were revealed and her case eventually reached the highest court, one of the most acclaimed jurists in American history, Supreme Court Justice Oliver Wendell Holmes, upheld the Virginia law and wrote, after reviewing her family history: "Three generations of imbeciles are enough."[1]

Investigations in recent years have revealed the true story behind the injustice done Buck. A Charlottesville native, she had been placed in foster care after the institutionalization of her mentally unstable mother. While in foster care, according to her account, she had been raped by a nephew of her adoptive family. When she became pregnant, her foster parents had her institutionalized as a "feeble-minded moral delinquent." She was then sterilized under Virginia's law authorizing this procedure for epileptics, the feeble-minded, imbeciles, and the "socially inadequate," whatever that might be determined to mean. Her infant was taken from her, to be reared by her adoptive family, perhaps out of the guilty recognition that their nephew had brought on the misfortune. The child died of an intestinal infection at age eight.

Several decades after the Supreme Court decision, researcher Paul Lombardo decided to look into the case. He discovered from school records that Carrie Buck had received better-than-average marks in both her lessons and deportment. There had been no evidence of mental deficiency. Lombardo was able to meet Buck shortly before her death in 1983 at age 76. She confirmed that the child born to her had been conceived during an assault. Still keenly feeling the shame of the events that followed and the shadow they had cast on her entire life, she especially regretted her inability to have children after her later marriage. In 2002, as a result of Lombardo's revelations, the Virginia legislature finally passed a resolution acknowledging the wrong done Carrie Buck. Sadly, there were no relatives left to savor the victory. The institution where Buck had been incarcerated is now the Central Virginia Training Center, providing services to people with mental disabilities. Buck's story is only one of the many tragedies resulting from eugenics policies. Most of the other stories remain untold.[2]

It was, however, the militarism and race policies of the German Nazis that definitively discredited social Darwinism and eugenics. In 1911, the German warmonger Friederich von Bernhardi published what would be remembered as a classic credo of German militarism. His *Germany and the Next War* touted the merits of warfare, which he found to be a law of nature, applying to humans as to other forms of life. The disciples of Bernhardi and those they influenced were among the villains who plunged the entire world into two devastating wars in which millions of people died miserably.

Early in their regime, the German Nazis set about to eliminate the mentally and physically deficient, along with many of the aged, in a program that aroused such an outcry in the population that it was finally disbanded. The Nazis did, as the world knows to its sorrow, succeed in declaring homosexuals, Jehovah's Witnesses, Gypsies, Jews, and other groups undesirable and were well on their way by the end of World War II to eliminating these populations in their horrible Holocaust.

NOTES

1. *Buck v. Bell* 274 U.S. 200 (1927). For a full account of the case, see Paul A. Lombardo, *Three Generations, No Imbeciles: Eugenics, the Supreme Court and Buck v. Bell* (Baltimore: Johns Hopkins University Press, 2006).

2. Andrew Pitzer, "Terrible Legacy of U.S. Eugenics," *USA Today,* June 24, 2009, 1–2B.

4

Theistic Evolutionists

Despite Richard Dawkins's assertion that Darwin paved the way for atheism, a surprising number of evolutionary scientists, who reject the approaches of both creationists and intelligent design proponents, still have been and are strong theists. Some have been members of traditional or even conservative religious bodies. Though Teilhard de Chardin may be the most celebrated of the theistic evolutionary scientists, he is by no means the only one. The group includes notable Europeans, North Americans, Australians, Asians, and probably others, with high scientific credentials and unquestioned integrity—some of the richest, most creative minds in the world.

INFLUENTIAL VOICES FROM WITHIN THE SCIENTIFIC COMMUNITY

Owen Gingerich, a professor emeritus of astronomy and the history of science at Harvard University, has long been associated with the Smithsonian Astrophysical Observatory. He is also a practicing member of the Mennonite Church. Although he believes in a universe with purpose and direction, he opposes the intelligent design movement, because he feels it unwisely and unfairly confuses science with religion. Gingerich's thinking is particularly influenced by the "anthropic principle," which acknowledges how unlikely intelligent life in the universe actually is, while marveling that it exists, despite all, on planet earth. To Gingerich, this hardly seems accidental.

Gingerich has dialogued with Philip Morrison on the possible existence of extraterrestrial intelligence. Morrison builds his own argument on antiquity, plenitude, and ubiquity. The universe is so old—over 10 billion years—that nature has had plenty of time to experiment with a vast variety of life forms. There are probably in the immensity of space a large number of habitable planets, Morrison contends. His decisive argument is the fact that the same laws of physics and chemistry that made life possible here on earth appear to operate elsewhere in the universe. While Gingerich remains skeptical, he refuses to be dogmatic, because that would mean putting a limit to God's creative power. Although he does not freely indulge in assertions of the miraculous, Gingerich feels that mutations might be more than blind chance events. Possibly they are instances of God's continuing involvement in the created world.

Father John C. Polkinghorne, an Anglican priest, is a former professor of mathematical physics at Cambridge University. He is an eloquent defender of the faith, its use of prayer and sacrament and its reliance on Divine Providence, and he sees no genuine conflict with modern science. Polkinghorne further believes that quantum theory has been helpful in dialogues between science and religion. Its revelation that the seemingly predictable world of everyday reality is "fitful and probabilistic at its constituent roots" suggests the flexibility that God has given himself to interact with his creation. Polkinghorne concludes that the laws of nature "do not constitute a straitjacket restraining divine action."[1]

Polkinghorne speaks of a God who is both being and becoming, though his commitment to "process theology" is limited.[2] Science, he teaches, has no right to any veto over theology, but theology must also strive for the fullest integration of all human knowledge. Polkinghorne, who is a pastor as well as a scientist-theologian, teaches that prayer is important, not that God needs to be informed of the needs and wishes of human beings but because humans need to be reminded of their constant need for divine assistance. There is no necessary choice between the God of the Bible and the God revealed in the pattern and structure of the physical world, according to Polkinghorne. Neither is it necessary to accept the clockwork, mechanistic world of Deism. The world revealed by modern science is open to "becoming," a world maintained by a living God who is constantly active in its process.

Acclaimed as a world-class geneticist, Francis S. Collins was the director of the Human Genome Project, one of the most important and successful scientific achievements of the 20th century, and has more recently become the director of the National Institutes of Health. In his spiritual autobiography, *The Language of God* (2006), Collins has described his conversion from atheism to evangelical Christianity, led by his scientific research and the writings of C. S. Lewis, the Anglican lay theologian, Renaissance scholar, and author of the *Chronicles of Narnia* books. Collins states emphatically that faith is never the enemy of scientific rationality. Faith and science, he believes, complete one another. God is actively at work in the world and is revealed by both science and faith. Still, as one whose work has opened the way for much further genetic study and research, Collins is sensitive to the contemporary fears of the misuse of science. He writes:

Is the science of genetics and genomics beginning to allow us to "play God"? That phrase is the one most commonly used by those expressing concern about these advances, even when the speaker is a nonbeliever. Clearly the concern would be lessened if we could count on human beings to play God as God does, with infinite love and benevolence. Our track record is not so good. . . . The need to succeed at these endeavors is just one more compelling reason why the current battles between the scientific and spiritual world-views need to be resolved—we desperately need both voices to be at the table, and not to be shouting at each other.[3]

Paul Davies, a cosmologist who also acknowledges the spiritual, has expressed his views in several writings, including *The Mind of God* (1995). He acknowledges that he is one of a group of scientists who does not subscribe to any conventional religion but still believes that the universe is purposeful. Although he does not suggest that the cosmos exists for humans alone, he believes that humans have a definite place in the scheme of things. He recognizes that many religious and metaphysical beliefs are childishly contrived and claim to know too much, while he believes we are barred from ultimate knowledge. However, he does entertain the possibility that the mystical experiences reported all over the world, from every religious tradition, may provide the only route beyond the limits to which science and philosophy can take us, "the only possible path to the Ultimate."[4]

The great scientist Theodosius Dobzhansky was another man of faith, a communicant in the Russian Orthodox Church. At the conclusion of *Man Evolving* (1969), Dobzhansky spoke highly of the aspirations of Teilhard de Chardin.[5] Teilhard, he reminded readers, saw the evolution of matter, life, and humanity itself as significant parts of a single process of cosmic development, a "coherent history of the whole universe." While Dobzhansky admitted that such a grand conception as that of Teilhard could not be demonstrated scientifically, neither was it a vision contradicted by secular knowledge. Teilhard was important in that he gave hope to modern humanity, seemingly so spiritually adrift in a meaningless universe. As one who maintained communion in the Eastern Christian faith, with its rich tradition of mystical yearning for God, Dobzhansky found the spirituality of the Jesuit priest highly congenial.

Kenneth R. Miller, biology professor at Brown University and author of numerous scientific papers and books, is a practicing Roman Catholic. He is also a strong opponent of intelligent design, testified against ID in the Dover, Pennsylvania trial, and is known for his humorous responses to favorite ID arguments.[6] Miller does not rule out God's continuing interaction with his creation. He feels the evolutionary process itself demonstrates both the humor and the limitless creativity of God. In his somewhat misleading but provocatively entitled book, *Finding Darwin's God* (2000), Miller includes his own credo. He quotes Darwin's own last paragraph in *On the Origin of Species*. Written before Darwin's surrender to agnosticism, it expresses wonder at the grandeur of the Creator, who from a few forms or one generated such an endless variety of wonderful and beautiful beings. This is the vision that captivates Miller.[7]

GERALD L. SCHROEDER

Gerald L. Schroeder is another scientist committed to strong, orthodox monotheism. He earned his PhD in physics from the Massachusetts Institute of Technology, where he taught for a number of years. Later he moved to Israel to work at the famous Weizmann Institute, where he continues his scientific research along with studies of the Hebrew Bible, the Talmud, and the Jewish mystical tradition. His books identify parallels between biblical doctrines and the research of biochemists, paleontologists, astrophysicists, and quantum physicists. He feels that religious belief is properly strengthened rather than shattered by an honest study of science, which, like the Bible itself, reveals the magnitude of the Creator. Cosmologists who have discovered the Big Bang and attempted to outline the beginnings of the universe and the inspired writers of the Bible who describe the first six days of Creation are dealing with the same realities, though in vastly different languages.

Schroeder contends in his book *Genesis and the Big Bang* (1992) that the biblical narrative makes clear that the plan of God was not to bring about a ready-made universe, as if by a stroke of magic. God chose a gradual unfolding, which is described in the first two chapters of Genesis.[8] Schroeder is not perturbed by the fossils collected by paleontologists; neither does he fear the discoveries of additional "missing links." He believes that many creatures, even a variety of hominids, preceded the arrival of Adam. Still, Adam's formation was of a nature different from the other objects and events of Creation. It was into Adam's nostrils alone that God breathed a living soul. All beings in the universe, organic or inorganic, Schroeder reminds readers, are composed of matter. Humankind is no different in physical nature. But to humankind alone is given the spark of divinity called soul. In this sense, Adam was indeed the first man. In a colloquial tone, Schroeder writes:

God might have plunked man down in a world that was readymade from the instant of creation. But that was not on the Creator's agenda. There was a sequence of events, a development in the world, which led to conditions suitable for man. This is evident from the literal text of Genesis l:l–31. By God's time frame, the sequence took six days. By our frame, it took billions of years. Regardless, there was a series of events separated by time. At the end of that sequence Man was formed.[9]

Schroeder believes the fine-tuning of the universe is further evidence of God's special plan for humanity. Like the fossil record, he finds the biblical description of life's unfolding to be punctuated. Each transition in the Bible—from nonliving to living, from plants to animals, from animals to man—is marked by a pronouncement of God. Schroeder is well aware of humanlike skeletal remains found in places like France and the Ukraine. These prove that creatures much like humans existed for the past million or so years. The physique of humans also changed as these millennia passed. But at a crucial juncture, which Schroeder dates at about 3700 B.C., a dramatic change took place. At this point, the partnership between God and humankind began.

For centuries, theological students have argued about the meaning of the "image" or likeness of God that the Bible says was bestowed upon humans. Schroeder believes that humans, who are certainly not made in a physical likeness of God, contain a touch of divinity. They are God's shadow, as it were, commanded to emulate him. This is probably what Christians also mean when they pray that God's will may be done "on earth as it is in heaven."

Schroeder appears to interpret much of the Bible literally, though he is always informed by a long tradition of rabbinical commentary and his rich knowledge of the connotations of Hebrew words. He speaks of the advanced ages attributed to certain worthies in the first books of the Hebrew Bible and believes these men of antiquity may actually have had very long lifespans. He has interesting things to say about Noah's flood, suggesting that it may have changed conditions on earth that had earlier favored longevity. He could no doubt have some rousing conversations with the Christian "flood geologists."[10]

T. O. SHANAVAS

Islam, another worldwide monotheistic faith, stresses the omnipotence of God and refuses to compromise his divine majesty. Unfortunately, most learned Islamic glosses on the Qur'an are not yet available in Western languages, and Islamic apologetic writings are not well known outside Muslim intellectual circles. A book that is accessible and does treat creation from an Islamic perspective is T. O. Shanavas's *Creation and/or Evolution* (2005). Shanavas was born in India and in 1970 immigrated to the United States, where he has been a practicing pediatrician. Though he is an applied scientist, his long interest in theoretical science and his wide scholarship have resulted in numerous articles in scientific journals. He is a member of the Islamic Center of Greater Toledo, Ohio, and is vice president of the Islamic Research Foundation in Louisville, Kentucky. His book argues that the Muslim view of creation is not incompatible with evolution, despite the contention of many fundamentalist Muslims in various parts of the world. He believes the Qur'an and the philosophers of Islam's medieval Golden Age provide full support for his contention that evolution itself is an intelligent design that manifests the Creator's power, supremacy, and dignity, while giving to humans limited free will that absolves the Creator of the wickedness of the world.[11]

Shanavas reviews some of the work of statesmen, jurists, historians, and scholars of the Islamic Middle Ages. He believes the scholars of this period anticipated some of the chief discoveries and conclusions of modern scientists. According to Shanavas, God knows all the directions and roads that the universe might take from the Big Bang to a final "Big Crunch." Yet God limits his omniscience and omnipotence in order to provide a measure of free will for his creatures:

The design of the universe, with the indeterminate nature of quantum behavior of basic matter along with the blended theological interpretations . . . allows us to conclude that Allah maintains His omnipotence, omniscience, love, and mercifulness without visibly violating any immutable laws of nature, allowing living beings to enjoy the gift of freedom.[12]

FRANCISCO JOSÉ AYALA

Two theistic evolutionists deserve special attention, because they have been especially active in disputes, have exerted enormous influence, and have written provocatively. The first is Francisco J. Ayala (1934–), who was born in Spain, became a Dominican priest in youth, and later immigrated to the United States, where he was the student and disciple of Theodosius Dobzansky. Ayala left the priesthood to devote himself fully to science and education, holding a special chair in the biological sciences, ecology, evolutionary biology, and philosophy at the University of California, Irvine. Frequently cited as one of the most eminent scientists still interested in religion, Ayala is known for the elegance and clarity of his presentations, whether writing or speaking.

Ayala opposes creationism and intelligent design, which he feels are based on fallacious science and promote a misunderstanding of religion. In 1982, he was an effective expert witness in Little Rock, Arkansas, against the attempts to introduce creationism into Arkansas public schools. Never one to mince words, he has accused the leaders of the intelligent design movement of "duplicity" and believes William Dembski, Michael Behe, and Phillip Johnson, among others, have been dishonest in their attempts to conceal the fact that they promote a particular religious point of view very thinly disguised as science. He points out that it is always the Judeo-Christian God that ID enthusiasts champion, though occasionally an antievolutionist appears who may suggest that the Designer could have been a space alien or a time-traveling cell biologist!

In response to the ID contention that evolution is just a theory and that a theory is not a fact, Ayala has offered a classic definition of scientific theory:

In science . . . a theory is a well-substantiated explanation of some aspect of the natural world that incorporates observations, facts, inferences, and tested hypotheses. Scientists sometimes use the word *theory* for tentative explanations that lack substantial supporting evidence. Such tentative explanations are more accurately called *hypotheses*.[13]

In answering ID arguments, Ayala asserts that the evolutionary origin of animals and plants is now accepted scientific fact, established beyond reasonable doubt. The evolutionary theory has been extremely beneficial to humans, essential in the development of highly productive crops that feed millions and in finding cures for devastating diseases. Evolution provides an explanation of why many pathogens that strike humans have developed resistance to drugs that were once effective. Evolutionary biology has explained relationships among wild and domesticated plants and between animals and their natural enemies. It has, further, been essential in understanding ecological problems and in seeking solutions to these problems. Instead of fighting evolutionary biology, religious people, Ayala believes, should honor the achievements of modern science, not only for its many benefits to humans but as a manifestation of the power and majesty of God.

Ayala is not impressed that the ID proponents have revived William Paley's argument from design. He feels that their evidence and arguments are bad science, with no genuine authority or coherence at all. Moreover, their arguments are bad theology. The design of organisms is not intelligent; the imperfections and dysfunctional features are evident for all to see. To attribute all this to the direct action of God, he feels, is incompatible with the assertions of the nature of God found in Christianity, Judaism, or Islam. Organs of the human body are less than perfect, modified from inherited structures rather than designed. Rudiments from earlier evolutionary forms are present in human embryos, including the brief emergence of a tail in the embryonic state. The eye, a favorite organ of the ID proponents, is far from perfect. It contains a blind spot, it is useless in the dark, and it is subject to numerous refractive errors. If it is so perfect, why do most people, after a certain age, require glasses to aid their vision?

Because of his distinguished work in several sciences and the brilliance of his speaking and writing style, Ayala has been called the Renaissance man of evolutionary biology. In his two books written for the intelligent layperson, *Darwin and Intelligent Design* (2006) and *Darwin's Gift to Science and Religion* (2007), Ayala has presented an argument not often heard in recent decades. He maintains that Darwinian theory has as much to offer religion as it does to science. He acknowledges that religion and science are separate disciplines, but he does not find them mutually exclusive. Science even offers some answers to a few of religion's most perplexing questions.

Christianity has long wrestled with the paradox of an omnipotent and all-loving God who allows so much suffering and evil to exist in the world. Mother Nature is an unbearably cruel parent. The forces of nature are not only heartless and mindless in their destructiveness; the cruelty of predatory beasts goes beyond any need for sustenance. Certain insects are cannibalistic; others devour their own mates, with some beheading their mates in a horrendous necrophilia before their mating has been consummated. Parasites kill and torment their hosts. Nature shows little compassion. Theological explanations for this paradox—that evil comes from the human abuse of free will and the original disobedience to the divine will—have never been totally satisfactory. Ayala feels that evolutionary theory provides the best resolution of this perplexity. Though God created the universe and the laws through which it operates, he is not directly responsible for the illnesses and violence that result from the secondary causes by which he allows the universe to operate.

Religious scholars in the past struggled hard to defend God's beneficence. How could he be responsible for the imperfections, dysfunctions, and the violence of the living world? Evolution, Ayala feels, has come to the rescue. It is therefore not surprising that Jack Haught, a contemporary Roman Catholic theologian, has written of "Darwin's gift to theology" or that the Protestant theologian Arthur Peacocke has referred to Darwin as the "disguised friend" or that Aubrey Moore, as early as 1891, wrote: "Darwinism appeared, and, under the guise of a foe did the work of a friend."[14]

Religious doctrines cannot be verified by science. Yet science, these theists assert, cannot be the only source of knowledge. Literature, art, philosophy, and religion also provide values and wisdom. Art and literature are not thought of as contradicting science; neither need religion be. The doctrines of the church—the Incarnation and Trinity, for example—are theological truths that can only be known through revelation. Science should leave them alone.

As a classical Christian, Ayala feels he is well within the thinking of the fathers of the Greek and Latin Church. He reminds his readers that Gregory of Nyssa (A.D. 335–394) as well as Saint Augustine maintained that not all species were individually created by God. Rather, some had evolved even in historical times. Gregory believed that the universe came into being in two successive stages. Stage one was instantaneous, while stage two unfolded gradually through time. Augustine was in accord with Gregory, teaching that many plant and animals species were created indirectly, in their potentiality, emerging through natural processes over time. In reviewing the writings of these church fathers and two thousand years of church history, Ayala feels he can speak with confidence of "Darwin's gift to religion."

TEILHARD DE CHARDIN (1881-1955)

The prince of theistic evolutionary scientists is certainly Father Teilhard de Chardin, not because his synthesis of science and religion is the most convincing but because of his originality, courage, eloquence of expression, and the grace of his personality. During his lifetime, his Jesuit order did not allow his most important writings to be published, and his work in paleontology was barely recognized in his own church. A Frenchman from an aristocratic family, he died in near-obscurity and was buried in the United States, far from his kin. Now, following the lessening of tensions between science and religion, the Roman Catholic Church's greater openness since the Second Vatican Council, and the generosity of spirit of Pope John Paul II, Teilhard is being recognized for his contribution to religion as well as to science.

Early in his career, Teilhard worked in England, where he was associated with the discovery and study of Piltdown Man, which turned out to be an embarrassing hoax. Stephen Jay Gould has suggested that the young Teilhard had a part in concocting Piltdown as a practical joke. Though this makes an interesting story, it is unlikely that Teilhard would have had anything to do with a fabrication that played directly into the hands of the opponents of evolution or further antagonized the officials of his church. His Jesuit superiors were constantly troubled by his views, which they believed bordered on heresy, and he was even for a time banned from participation in the life of his order.

Despite his Jesuit loyalties, Father Teilhard largely followed an independent path. There were a number of important women in his life, though all these relationships were almost certainly of a nonphysical nature. These women, however, were a curious lot for a practicing Jesuit priest to be so strongly attached to. There was Ida Treat, a pro-Communist political activist, an American, and a scientist. Another was Leontine Zana, a Catholic but independent minded, philosophically inclined, and a feminist. Still another, Jeanne Mortier, was a student of Teilhard's thought, whom he appointed to oversee some of his posthumous literary publications. In later years, he appears to have been especially dependent on the wife of a fellow scientist, the novelist Rhoda de Terra. The women in his life often identified with him so profoundly that strong jealousies emerged among them.

But the most important woman in Teilhard's adult life was certainly Lucile Swan, a Chicago divorcée and artist, whose long, platonic, but decidedly romantic attachment to him started in China and continued with visits in Paris and the United States. In his frequent letters to Swan, Teilhard outlined many of his central ideas. She also read his unpublished work, largely concurred with his thought, and translated some of his writing into English. Though not herself Roman Catholic, she was deeply spiritual and of a mystical cast of mind. She would have preferred a more conventional male-female relationship, which Teilhard avoided by reminding her of his priestly vows. This was no doubt a ready excuse; he appears to have been so absorbed in his work that romantic complications would have been avoided even had he been a layman. Still, Teilhard was the center of Swan's life, and after his death, she comforted herself with Eastern spiritual techniques and philosophies.

The letters of Teilhard and Swan are interesting, because they outline events in the priest's life on several continents as well as his teachings in concise form. In these letters, as in many of his writings, Teilhard seems hardly to distinguish between his scientific discoveries as a paleontological field worker and his religious ideas. Everywhere he seeks to achieve a unity—almost in an Eastern religious fashion—and he talks of human relationships and everything else as somehow "converging," one of his favorite words and concepts. To Swan, her beloved Teilhard always dwelt on a superior plane, even as she lamented the aloofness she found in him and her inability to remain on this plane herself.

In a letter to Swan from Peking (Beijing), dated January 25, 1937, Teilhard spoke of his passionate love for the world, even in a pagan sense, he admitted. In this respect, he could never have been a true mystic in the Eastern way, one who detaches himself from the material world. But he professed an equally intense devotion to the God who expresses himself in Jesus Christ. Though he acknowledged his love for Christ to be the product of his education, he credited this love with saving him from falling into a pagan pantheism, to which his spirit naturally inclined. It was Christianity that had imparted to him an "incomparable adoration" for a Person rather than some vague Ground of All Being. Confronted with Christ, he perceived less the Restorer of a fallen world than the Animator of a "universe in progress." The essence of the Gospel, he believed, is the expression of a constant aspiration to achieve new levels of being and perception.[15]

On March 24, 1947, Teilhard wrote Swan from Paris that, despite opposition from his order, he would never totally let up. Although many existentialists and pessimistic Christians could not understand him, he felt his own position growing stronger and clearer in his contemplations and in his discourses with others. He saw humanity, while not yet perfected, moving toward some great future goal. Any religion that satisfied men and women had to incorporate this hope.

Teilhard's most important work, in the view of his devotees, is *The Phenomenon of Man*, published posthumously in 1955, with an American edition in 1959. Although more and more theistic scientists have read this book with enthusiasm and even a pope has acknowledged its worth, there is still the open question of whether Teilhard is truly within the mainstream Christian tradition. Some of his most conscientious readers find his thinking more congenial to Hinduism than to Christianity.

In his introduction to the English translation of *The Phenomenon of Man*, Sir Julian Huxley, an atheist who was one of Teilhard's strongest supporters and friends, proclaimed the book "a very remarkable work by a very remarkable human being." The introduction went on to summarize succinctly the leading ideas of the book. Teilhard accepted evolution as self-evident; he viewed human beings as natural phenomena, proper objects for scientific study, as are all material beings. He saw the entire universe as a gigantic process, a process of becoming, constantly striving for and attaining new levels of being and integrity. Consequently, it became appropriate for him to speak of *cosmogenesis* rather than simply cosmology. Teilhard preferred the term *hominisation* for the process by which he believed humans moved forward, achieving more and more of their potential along the way. Human evolution, just as the evolution of all else in creation, had to be considered not chiefly as a matter of origin but a matter of direction, toward the achievement of inherent possibilities.

Even Huxley, one of the keenest students of Teilhard's thought, admitted to ambiguities and occasional obscurities. Like numerous other scientists, he was unable to follow Teilhard in "his gallant attempt to reconcile the supernatural elements of Christianity with the facts and implications of evolution."[16] For the English reader of *The Phenomenon of Man*, there are additional problems; the obscurities remain, along with a certain muddiness in the English translations.

As he looked from the human past into his envisioned future, Teilhard spoke of a process of convergence toward a final state, which he called the "Omega Point," as opposed to the alpha of elementary material particles and their energies that existed at the beginning. As knowledge increased, he believed that human individuals and societies would achieve a hyper personal relationship with an emergent Divinity, though it is not totally clear just what this entails or the nature of this emerging Divine Force. Still attempting to anchor his thought within Christian tradition, Teilhard spoke of the process as *Christogenesis*.

Teilhard's fundamental propositions were relatively few. He believed that all phenomena—the cosmos, all organisms, human beings—should be viewed as constantly dynamic and emerging in an evolutionary manner within space and time. Through humanity, the universe will evolve to become eternally concentrated at the Omega Point, which will then be immaterial and free from mutability. This process and purpose of evolution has been planned by God, who is himself the Omega into which all consciousness will finally be concentrated. Though his thought sounds increasingly pantheistic, Teilhard clearly intended this Omega Point to be identified with the personal Christian God.

Despite his renown, his welcome by scientists who also want to be spiritual, and his personal dignity, Teilhard and his work have occasioned severe criticism, and not just from within his church, where he was first challenged. It was, after all, not so much his belief in evolution that caused his own order to suspect his orthodoxy; it was his alleged weakening or rejection of the doctrine of original sin. While today the Roman Catholic Church is ready to acclaim Teilhard, the chief attack on his work now comes from scientists. Over and over, scientists and science educators remind creationists and ID advocates that science and religion inhabit different, nonoverlapping provinces. Each must respect the territory of the other. Yet Teilhard presented his philosophical-religious meditations as scientific writing. There is, of course, no valid way his philosophical-religious hypotheses could ever be tested by the scientific method.

P. B. Medawar wrote a highly critical review of *The Phenomenon of Man* shortly after it appeared in English.[17] This review, frequently reprinted but originally published in *Mind*, clearly identifies the difficulties in Teilhard's thought. Medawar first rejects the extravagant praise of numerous French readers who proclaimed *Phenomenon* the book of the year or even the book of the century. On the contrary, Medawar finds in it a feeble argument, poorly expressed. He pronounces its prose-poetry, so admired in the French edition, "tipsy, euphoric," characteristic of "the more tiresome manifestations of the French spirit." He quickly identifies errors of fact and contestable judgments, which Teilhard expressed in an extravagant style filled with abundant neologisms.

Despite the muddle of words, Medawar acknowledges that it was still possible to discern a train of thought in *Phenomenon*. This is Teilhard's belief that the fundamental process in the entire universe is evolution. Though unsupported by further scientific evidence, Teilhard goes on to insist that evolution "has a main track or privileged axis." All the universe moves toward consciousness, with present human consciousness still advancing toward some culmination in supreme consciousness, designated the Omega Point, which assimilates into itself all personal consciousness. Medawar clearly understands that this Omega is identified with God, because at one point in his writing, Teilhard even refers to a "God-Omega." Everywhere, souls are breaking away, carrying upward their silent load of consciousness, Teilhard continues, ascending to that great collectivity of consciousness.

Medawar marvels at the enormous following for what he regards as Teilhard's mystical mumblings. He calls *Phenomenon* a sort of "philosophy fiction" rather than anything that approaches legitimate science. And he identifies the likely audience as a large group of people who have been formally educated far beyond their capacity for analytical thought. Medawar goes on to suggest the reasons for Teilhard's contemporary popularity. First, it is a basically antiscientific admiration for a man with a scientific education who achieved a moderate success in paleontology, not the most exacting of the sciences. Second, Teilhard's "totally unintelligible style" substituted for profundity in the eyes of many readers. Third, Teilhard addressed the human situation, felt to be especially grim after the European experience of World War II, and he appeared to present a remedy. Finally, and this is especially important, Teilhard was introduced to the English-speaking world by Sir Julian Huxley, heir to the eminent British family inseparably associated with Darwin. As the grandson of Darwin's "bulldog," Thomas Henry Huxley, Sir Julian gave his imprimatur to the work. With such effective sponsorship, Teilhard's writing has been taken seriously; it appears scientific because it is tied to a vague, general conception of evolution, though far from Darwin's own. Because people readily believe what alleviates their anxieties and strongly desire a scientific validation of the religion that has sustained them in the past, they are willing to accept with relish what Medawar calls Teilhard's "bag of tricks."

Despite powerful critics such as Medawar, with every year Teilhard de Chardin seems to attract new disciples, along with critical evaluations of his work. His writing is certain to attract attention for some years to come, and the attractiveness of his personality is evident as new biographies and collections of his essays and letters appear.

NOTES

1. John Polkinghorne, *Science and Providence* (West Conshohocken, PA: Templeton Press, 2005), 30.

2. Process theology stems from the philosophy of Alfred North Whitehead, expanded and developed by several 20th-century theologians. It teaches that the universe is characterized by process and change. God interacts with this changing universe and therefore has a changing element in his own nature, though his goodness and wisdom remain eternally the same. All religion and scripture are human interpretations of divine action.

3. Francis S. Collins, *The Language of God* (New York: Free Press, 2006), 272.

4. Paul Davies, *The Mind of God* (New York: Simon & Schuster, 1995), 232.

5. Theodosius Dobzhansky, *Man Evolving* (New York: Bantam Books, 1969), 364.

6. Kenneth R. Miller, *Only a Theory: Evolution and the Battle for America's Soul* (New York: Harper Perennial, 2008), 53–57.

7. Kenneth R. Miller, *Finding Darwin's God* (New York: Harper Perennial, 2000), 292.

8. Gerald L. Schroeder, *Genesis and the Big Bang* (New York: Bantam Books, 1992), 48.

9. Ibid., 85.

10. Ibid., 32.

11. T. O. Shanavas. *Creation and/or Evolution* (Philadelphia: Xlibris, 2005), 194–195, 210, 212, 216–217.

12. Ibid., 218.

13. Francisco J. Ayala, *Darwin and Intelligent Design* (Minneapolis: Fortress Press, 2006), 73.

14. Ibid., 89.

15. Thomas M. King and Mary Wood Gilbert, eds., *The Letters of Teilhard de Chardin and Lucile Swann* (Washington, DC: Georgetown University Press, 1993), 65.

16. A good summary of Tielhard's thought may be found in the essays on his mysticism found in Philip Appleman's edition of *Darwin* (New York: W.W. Norton, 1970), 459–485.

17. See P. B. Medawar's critical review of *The Phenomenon of Man* reprinted in Appleman's *Darwin,* pp. 476–485.

5

Arguments for Creationism and Intelligent Design

Evangelical religion was a dominant force in U.S. life in the middle of the 19th century. Despite the constitutional separation of church and state, almost unique in the late 18th century, Americans remained a distinctively religious people. Doctrinal disputes about the second coming of Christ and varying interpretations of "end times," appealing to the apocalyptic biblical books of Daniel and of Revelation, were matters of prime concern in the last decades of the century. Controversies centered on the thousand-year reign of Christ on earth that these books were believed to predict. Premillennialists, who believed Christ would return to rescue earth during a thousand years of peace, viewed present-day culture as sinful. Postmillennialists, on the other hand, who believed Christ would appear after a thousand years of peace, accepted the present age and expressed a more optimistic view of secular culture. The former were more inclined to see nature and the entire world as a "tooth and claw" struggle, while the latter would find the lack of progressive vision in Darwinism unacceptable.

Liberal theology, which developed among an intellectual elite in Germany, made some inroads near the end of the century, especially in denominations with seminary-trained clergy. Liberal theologians viewed the Bible as an ancient book of wisdom, at best containing valuable insights into divine-human relationships and ethical conduct but no more reliable in matters of history and geology than other ancient writings. Scholars observed that ancient documents, biblical and otherwise, habitually mingled factual history with accounts of the miraculous. Accepting the Bible as written, ordinary laypeople paid little heed to liberal views, at least as long as their ministers kept them out of the pulpit.

YOUNG EARTH CREATIONISM

In the early 20th century, Evangelicals showed some tolerance of modified evolutionary ideas. Darwin had a number of now largely forgotten defenders among them. Some were ready to interpret the Bible in a manner that permitted a long history of the earth. Still accepting the Genesis account of creation, the "gap theory" was advanced, allowing enormous periods of time to unfold between each of the seven days mentioned in Genesis. The "day-age theory" also made most of the findings of geology acceptable. Because the Bible teaches that one day is as a thousand years in the eye of God, a Genesis day could represent any length of time, it was reasoned. There was even a school of thought that accepted the existence of pre-Adamic hominids, affirming only that Adam and Eve were the first fully human creatures on whom God bestowed his image.

In the 1920s, fundamentalism emerged as an important emphasis in American Christianity. Very numerous, fundamentalists revolted against both Darwinism and German liberal theology, affirming that the Bible was not only an infallible guide to faith and morals but inerrant as well in its history and geology. The creation science movement emerged from these concerns. The first important leader in this renewed movement of biblical literalism, in confrontation with science, was George McCready Price, whose most popular book, *The New Geology* (1923), propounded a "new catastrophism" whose centerpiece was the Genesis flood, viewed as the major geological event in the history of the world, accounting for the fossil record and the disturbances in mineral deposits that geologists had documented. Although a quick study and a studious man deeply interested in science, McCready lacked the sophisticated training in geology that he might have received had he been able to attend a major university. Still, he was determined to establish empirical proofs of his young-earth beliefs that would then be verified by geology and paleontology. *The New Geology* was designed as a textbook to be used in classrooms. In it, Price rejected both an ancient earth and Darwinism. The Bible was his touchstone against which all knowledge must be tested.

Creation science, developing from the ideas of Price, Duane Gish, and others, has had an enormous popular following. As it sought inroads into the public schools, it frequently downplayed its biblical objections to evolution and attempted to present its ideas in scientific language. No longer seeking, as had earlier antievolutionists, to ban Darwin from the classroom, creationists now asked instead that their views be presented alongside evolutionary approaches in science classes. They believed that evolutionists in their popular writings had ignored reasonable questions and genuine problems with their theories. These questions had to be addressed. For example, where were the missing links to species? How could the Cambrian fossil explosion, in which an enormous variety of forms seemed suddenly to appear, be explained? Why had no evolutions of new species occurred even in laboratories where insects with short life and reproduction spans were closely observed?

Despite attempts to anchor creationism in science, the movement failed in the courts, which branded it a sectarian religion thinly disguised by scientific jargon. The doctrines of creationism then retreated into private schools and Bible colleges supported by conservative religious groups. But its faithful have not relinquished their beliefs. Instead, they have developed a rich counter-culture, with their own presses, numerous publications, films, and museums.

In his chatty and amusing book *Rapture Ready!* (2008), Daniel Radosh has described his visits to creation museums. The highlight of his little odyssey was his trip to the Answers in Genesis Creation Museum in northern Kentucky. A 27-million-dollar structure that opened in 2007, this glitzy operation is the showcase institution of young-earth culture. The museum, as Radosh describes it, spreads out to 70,000 square feet of state-of-the art exhibits, the work of designers with elite clients such as Universal Studios. There are walk-through jungle dioramas, waterfalls running over fiberglass rocks, and tanks filled with fish and turtles. Dark-skinned mannequins appear beside small dinosaurs. "Long ago, dinosaurs and people were friends," one exhibit proclaims, promoting the ultimate Bambism. In another section of the museum, children are allowed to sit in a leather saddle atop a model of a triceratops. A museum guide explains that it is quite likely that humans once were able to domesticate dinosaurs.[1]

In his research, Radosh also learned that Americans spent at least 22 million dollars in 2004 on creationist lectures, academic texts, and other pop culture artifacts promoting these beliefs. Much of this was spent on books and toys designed for children, including picture books, coloring books, biblical comics, and tales of the Piltdown hoax, which give the impression that scientific circles are given to much misrepresentation. Some of the books for adults discuss UFO sightings and accounts of alien abduction, suggesting that these are demonic attacks.[2]

When questioned about the abundance of dinosaurs in creation museums and publications, curators acknowledged that the appeal is chiefly to children, who are always fascinated by these creatures. Children, along with their parents, come in great numbers to the museums, primarily for the dinosaurs, which admittedly are not mentioned, at least directly, in the Bible. As he watched the busloads of children from Sunday schools and home schools arrive at the museum, spend hours at its exhibits, and take home souvenirs from the gift shop, Radosh concluded that creationism is more than just a religious movement; it has taken its competitive place in U.S. popular culture. And what Americans do as well as, if not better than, any other people in the contemporary world is entertain!

INTELLIGENT DESIGN

At the same time young earth creationism was forming its own culture, a more sophisticated challenge to Darwinism appeared, now promoted not so much by clergy and religious laypeople as by legal scholars and people with science doctorates from major universities. Whatever the personal religious views of its champions, the intelligent design movement, as it came to be known, carefully avoided words with religious connotations, did not openly anchor its assertions in the authority of the Bible, and attempted to present arguments that would appeal to people instructed in modern science.

The simplest definition of ID is that of the Discovery Institute of Seattle, Washington, its prime think tank. ID is a theory that some things are "best explained by an intelligent cause."[3] Because nobody was present at Creation, its events are impossible to verify according to the scientific method. But this problem is faced by other cosmologists, biologists, and chemists who are themselves given to speculations that cannot be tested according to the scientific method, at the same time they condemn the propositions of the ID movement.

Although ID has been a blanket movement that has not sought to exclude any believers in a Prime Mover, its strongest spokespersons do not accept the young earth creationism of a McCready. Most of them concede that the earth is millions of years old, and many believe that at least some microevolution has taken place. But they do have problems with Darwinism and identify several issues that have not yet been adequately addressed by evolutionists.

The designation *intelligent design* only goes back to 1988, when it was first used by Charles Thaxton in a speech, and later in the textbook he edited, *Of Pandas and People: The Central Question of Biological Origins*. Many consider Phillip Johnson the real founder of the modern ID movement. Though not a scientist, Johnson is one the United States' most distinguished jurists. He started questioning the integrity of the scientific establishment on a visit to the British National History Museum in 1988. At that time, he learned that a display by the museum's paleontologists had been removed because it had presented Darwinism as "one possible explanation" of human origins. Because Darwinism appeared to be questioned, influential persons in the science community had denounced the exhibit and forced its cancellation. An enemy of political correctness, which had taken over American university campuses, often dictating the way humanities and social studies were being taught, Johnson now saw the same forces operating in the hard sciences, shutting off legitimate dissent and enforcing ideological conformity.

Coming rather late to a concern with evolution, Johnson gave his attention to Darwinian theory, which he found flimsy, riddled with uncertainties and gaps. Evolutionary scientists, he felt, often made wistful jumps, reaching conclusions from scant evidence. A sudden appearance in the fossil record would be described as "rapid evolutionary branching"; slight variations in moths or fruit flies would be heralded as "macro evolutionary evidence"; while elaborate models of supposed human ancestors would be constructed from a few bone fragments. "Missing links" would be wildly hypothesized. Evidence of this sort, Johnson pointed out, would be quickly dismissed in any U.S. court of law.

For any thoroughgoing evolutionary system, Johnson also concluded, atheism was practically a necessity. Yet commitments to atheism were made even before data were examined, and interpretations of such data were themselves based on atheistic propositions. In his 1991 book *Darwin on Trial*, Johnson presented his conclusions in the clear, logical fashion of a master attorney and met with an appreciative response. The ID movement was soon under way, as PhDs in several scientific disciplines and others with reservations about mainstream evolutionary science came together in the Discovery Institute. The board members and fellows soon included people of several religious and philosophical backgrounds: Protestant, Roman Catholic, Eastern Orthodox, Jewish, and even one well-known agnostic.

Despite the ecumenical composition of the Discovery Institute and its attempts to avoid appeals to religious authority, ID champions soon discovered that their papers were not welcome in leading refereed scientific journals. Even with the highest degrees, successful teaching experience, and respectable records of research, they were frequently denied tenure and promotion in the institutions where they taught. If they managed to obtain tenured faculty positions, despite all this, their departmental colleagues and sometimes the institutions themselves might issue formal statements disavowing their views. It was not surprising that Discovery Institute fellows suspected a conspiracy against them and protested that their academic freedom and basic freedoms of speech and press were being challenged. The large sales of their books and the constant demand for their services as public speakers were a partial compensation for their lack of acceptance in their individual scientific disciplines, and the opposition to their ideas merely confirmed their faith in their validity.

Despite the diversity of leading ID advocates, their common affirmations can be summarized in the following manner. First, they find the universe too complex and too hospitable to life, particularly human life, for it to have come into being through blind energies. The universe is precisely calibrated to make possible the emergence of rational, introspective life. The fact that the existence of an Intelligent Designer eludes scientific scrutiny proves nothing. Some of the leading evolutionists—E. O. Wilson, Richard Dawkins, Daniel Dennett—are openly and aggressively atheistic, yet atheism, like theism, refuses to yield to scientific verification. The fact that evolutionary atheism has itself become a religion is a favorite theme of the Discovery Institute. Because ID is emphatically not concerned with biblical accounts of Creation but looks for its support in science itself, it is as scientific as atheistic evolution, or so its proponents contend.

The arguments of ID enthusiasts are many. The concept of a Designer, First Cause, or Creator, they point out, has been accepted by most people throughout history. ID does not try to identify this Designer with the God of a particular religion, though individuals within the movement are free to form their own theological views. ID further recognizes that religion has been the germinating force for the development of the sciences. Almost all the great scientists of the past built their theories on the acknowledgement of a Designer. Religions and science need not be in conflict in modern times, just as they were not in ancient times.

Most ID proponents acknowledge that microevolution has been established through scientific research. Small changes do take place in species; laboratory observations of moths, fruit flies, and other insects that reproduce rapidly have established this. Selective breeding among domestic animals further affirms small changes. However, macroevolution is more problematic, especially when it applies to the evolution of humans from lower forms of life. When, they ask, has the evolution of one species from another been observed?

At the very least, Darwinism and the neo-Darwin synthesis need major corrections; there are still too many gaps. According to some mathematical calculations, there has not been enough time since the beginning of the universe—as contemporary science reveals it—for human evolution, as the Darwinians describe it, to have taken place. Evolutionary theory also seems to violate the Second Law of Thermodynamics, which states that left alone, a system will always move from relative order to disorder. The concept of random mutation, so important to evolutionary theory, is far from established. Neither can natural selection adequately account for the enormous diversity of life. When Louis Pasteur's work debunked "abiogenes," demonstrating that spontaneous generation cannot take place, he would seem to have also established that living beings cannot come from nonorganic material. Neither has the problem of infinite regression been solved by the evolutionists.

At the time he was presenting his theories, Darwin conceded that the fossil record was still sketchy. He was nevertheless certain that in time this record would be fleshed out, and paleontologists have constantly made discoveries that appear to do just that. Still, ID scholars argue that the lack of transitional fossils remains a problem. They point to a few highly publicized hoaxes such as the Piltdown Man and the tendency of researchers to extrapolate too much from a few questionable bones that may be found in ancient bedrocks. The Cambrian explosion, with its sudden appearance of the fossils of numerous species, continues to present other problems for traditional evolutionists, and ID advocates are not ready to simply accept the explanations of Stephen Jay Gould and Niles Eldridge that evolution operates in spurts of activity.

One of the favorite ID arguments centers on "irreducible complexity." Michael Behe has elucidated this concept more forcefully than anyone else, and it is basic to his argument in *Darwin's Black Box* (1996). He defines the term as follows:

By irreducible complexity I mean a single system composed of several well-matched, interacting parts that contribute to the basic function, wherein the removal of any of the parts causes the system to effectively cease functioning. An irreducibly complex system cannot be produced directly {that is, by continuously improving the initial function, which continues to work by the same mechanism} by slight, successive modifications of a precursor system, because any precursor to an irreducibly complex system that is missing a part is by definition nonfunctional.[4]

Favorite examples of such irreducible complexity that Behe uses are the blood clotting system, the human eye, and the bacterial flagellum.

Physics is the scientific discipline most congenial to ID. The Big Bang explanation of the beginnings of the universe is especially welcome to ID champions, particular those who believe the universe was created ex nihilo. But biologists and chemists within the ID community find other arguments in support of their belief that life is too complicated to be explained satisfactorily and fully by evolutionary theories. "Specified complexity" is a concept carefully developed by William Dembski, who believes it is not enough for an organism to be complex to prove design. It must also have an accomplished purpose that could not have been achieved by random forces. He frequently uses an example from the movie *Contact*. The scenario of the film was based on a premise of the SETI (Search for Extraterrestrial Intelligence) project. In the film, a complex radio signal is received from outer space, a series of beeps and pauses that correspond to a sequence of prime numbers, indicating specified complexity, which can only be a message sent by intelligent beings.[5] Dembski believes there are numerous examples of such complexity to be observed in the material world.

ID, despite its careful avoidance of religious language and its efforts to ground its arguments on solid science, has fared little better in the courts than the earlier creation science. Textbooks favorable to ID have also been driven from the public schools. ID advocates believe this is the result of an academic orthodoxy that is stifling open inquiry and any difference of opinion. Political correctness, they feel, has invaded the universities and professions and is not favorable to the work of ID scientists.

Because Darwinism's conflict with scientific creationism and intelligent design has been as much a matter of clashing personalities as opposing philosophical systems, it is instructive to examine a few of the leading personalities of the movements.

CREATIONISM AND DUANE TALBERT GISH (1921–)

Scientific creationism has not been totally swallowed by ID. Although it has been decisively driven from the public schools, scientific creationism retains a substantial following in fundamentalist, independent, and Pentecostal churches. The most prominent figure in the movement since the time of George McCready Price is Duane Talbert Gish, an energetic lecturer, debater, and writer, known for his dramatic performances and cantankerous personality. A former vice president of the Institute for Creation Research, Gish has been facetiously called "the Thomas Henry Huxley of Creationism." He is a trained scientist, with a PhD in biochemistry from the University of California, Berkeley. During his time at Berkeley, he was an assistant research associate, and he later served as an assistant professor at Cornell University Medical College (now Weill Cornell Medical College), where he conducted biomedical and biochemical research for about eighteen years. He also worked for a time as a research associate for the Upjohn Company.

But, through it all, he has been a student of the Bible. Although reared in Methodism, he later joined the Baptists and became an ardent defender of the full literal integrity of the scriptures. During his career as a researcher, he also became convinced that scientists were falsifying evidence in an attempt to undermine the Bible. By the 1960s, Gish had become very active in the antievolution crusade, lending his talents to its organizations. During the 1970s, he joined the faculty of San Diego Christian College, a school committed to literal biblical interpretation. Shortly thereafter, he accepted a position with the Institute for Creation Research, though it is not clear what results his research with this organization produced, other than his array of books and lectures. The last position he held was senior vice president emeritus with that organization.

Though born in Kansas, Gish received his higher education in California and later settled in Dallas, Texas, an especially hospitable environment for his teachings. Accepting speaking engagements throughout the continent, he quickly became known as a lively debater, with a shotgun style of delivery and a habit of quickly shifting topics. His detractors referred to his style as the "Gish Gallop." Because he rarely deviated from his stock speeches or added new debating points, his opponents soon learned what to expect from him. His asides and jokes also remained the same in debate after debate, even as his audiences never seemed to tire of them. Detractors accused him of concocting his facts and figures, as well as misquoting both his sources and his opponents. Ignoring any attempt to correct his facts, he often responded to criticism with personal attacks on these opponents. But because Gish usually lectured to audiences in churches or to others who already agreed with his views, he could constantly claim success. Most scientists who confronted him soon concluded that it was useless to debate him, that their presence on any platform with him only increased his standing and gave credibility to what they believed were his weak arguments. Nevertheless, he has shared the stage with a number of noted personalities, including Ian Plimer, head of the geology department at the University of Newcastle in Australia.[6]

Gish has never been hesitant to answer his critics. Joyce Arthur published a thoroughgoing criticism of him in 1996, in *Skeptic,* the publication of the Skeptic Society. Gish's response the following year also appeared in *Skeptic* and reiterated the main talking points of his lectures.[7] The modern creationist movement, he argued, was by no means attempting to introduce the biblical narrative into public schools by disguising it as science. The only intention of creation scientists, he contended, is to make scientific evidence for creation available to schoolchildren so that they may form their own judgments. The means to that end include an examination of the fossil record, the laws of probability, the laws of thermodynamics, the evidence of purpose and design in biology, and the vast evidence from other areas of science that provides proof that living organisms were created by an Intelligent Agent external to the natural universe. He further stated that his program presents a choice for students between a special creation by supernatural power and a mechanistic, naturalistic evolutionary origin of living beings and the universe. Scientific creationists, he reiterated, intend that this be done without reference to the Bible, any devotional literature, or humanistic manifestos.

Gish viewed both creationism and evolution alike as historical rather than scientific theories. But quite apart from the merits of the contrasting theories, Gish reminded his audience that the overwhelming majority of U.S.

taxpayers, who finance the public schools, want children exposed to both points of view.

In further response to his critics, Gish claimed, even while acknowledging human fallibility, that he had never written or spoken in lectures anything that he knew to be false. He also denied ridiculing or slandering other scientists. It was they, in fact, who had distorted the views of creation scientists, lumping all of them together and labeling them a fringe group. He was particularly offended by the skeptics who accused creationists of paranormal claims, attending séances, and reporting UFO sightings. Attempts to associate creationists with cults, he alleged, was an unethical effort to divert attention from the mass of evidence, both historical and scientific, supporting a supernatural origin of the universe. Creationism, he concluded, is not only essential to understanding science but vital to the eternal destiny of each human being.

INTELLIGENT DESIGN LEADERS

Intelligent design beliefs have proved attractive to many people who consider themselves more scientifically informed than earlier opponents of Darwinism. The leaders of the movement are celebrities within their own circles, their books and speeches in constant demand. The most visible members of the movement, with their impressive academic degrees, are active debaters, always challenging their opponents in forums mostly attended by confirmed partisans of ID. These audiences are certain that evolutionists have been demolished by the learning or logic of men like Stephen C. Meyer, Phillip E. Johnson, Michael J. Behe, William A. Dembski, and others.

STEPHEN C. MEYER (1958–)

Stephen C. Meyer is widely recognized as the cofounder, along with legal scholar Phillip E. Johnson, of the intelligent design movement. A senior fellow of the Discovery Institute and director of the Center for Science and Culture at the Discovery Institute in Seattle, Washington, he holds a PhD in the history and philosophy of science from Cambridge University and has made special study of scientific research methodology and the history of biology. At one time, he was employed by Atlantic Richfield as a geophysicist. His writings have been published by both religiously affiliated presses and by the Michigan State University Press.

Meyer, who is a Presbyterian, taught for a while in a Christian university but left teaching to devote himself full-time to the intelligent design movement. In 1999, he and associates mapped out the plan, outlined in the Wedge

Document, for introducing ID into public schools.[8] Although it met with a good initial reception, largely because of the enthusiasm of parents, opponents of ID have regarded the Wedge as a subversive plan of action.

In August 2004, one of Meyer's articles appeared in the peer-reviewed scientific journal, *Proceedings of the Biological Society of Washington*. It occasioned such an uproar that the journal's publisher, the Council of the Biological Society of Washington, later disavowed it, claiming it had not met the scientific criteria demanded by the publication and denying that it had been properly peer reviewed. The managing editor of the journal also came under attack. Because of so many similar experiences, the ID community continues to protest its ostracism by the scientific establishment. Supporters of Meyer point out the increasing challenges to employment and academic tenure, along with further forms of intimidation, including threats of violence to scholars associated with the ID movement. These mounting actions continue to be interpreted as a conspiracy to suppress dissent.

MICHAEL J. BEHE (1952–)

Another senior fellow of the Discovery Institute, who is much in the public eye with writings, lectures, and appearances as expert witness in court cases, is the biochemist Michael J. Behe. His PhD is from the University of Pennsylvania, and his chief research interest has been the delineation of design and natural selection in protein structures. He claims some 35 articles published in refereed journals (possibly before he became widely known as a Discovery Institute fellow), in addition to his popular essays for newspapers and periodicals. Though he holds a tenured professorship at Lehigh University in Pennsylvania, the official Web site of his department contains the following statement:

While we respect Prof. Behe's right to express his views, they are his alone and in no way are endorsed by the department. It is our collective position that intelligent design has no basis in science, has not been tested experimentally, and should not be regarded as scientific.[9]

Behe's admirers, who are many, have pointed out that few universities around the country have seen fit to publicly disclaim the many controversial and often highly eccentric social and political views held by their professors. And they ask why this well-qualified professor should be so singled out.

Despite his own colleagues' disenchantment, Behe's highly technical books have attracted international attention, sold well, occasioned much discussion, and even been pronounced by *National Review* and *World* magazines as among the 100 most important books of the 20th century. In addition to his technical writings, Behe has been skilled in popularizing his ideas in editorials for such publications as *Boston Review, American Spectator,* and the *New York Times.* Along with his associates at the Discovery Institute, William A. Dembski and David Berlinski, he is credited with tutoring the popular writer and television personality Ann Coulter for the science sections of her book *Godless: The Church of Liberalism* (2006), a best-seller read by millions. Although he does not present his ID views as religious, and he is neither denounced nor endorsed by his church, Behe is a practicing Roman Catholic layman.

His central idea, which he did not originate but which he has fully explored, is that of irreducible complexity, with his familiar examples of blood clotting, the human eye, and the bacterial flagellum. Some have identified Behe's argument as a modernized version of William Paley's watchmaker thesis. Behe appears to accept limited evolution (microevolution), which he feels can explain marginal changes in the history of organisms. But he argues that most of the mutations that have brought changes to life on earth have not been random, as Darwin thought.

WILLIAM A. DEMBSKI (1960–)

Still another high-profile ID proponent is William A. Dembski. He holds a PhD in mathematics from the University of Chicago and a PhD in philosophy from the University of Illinois, with subsequent degrees in theology and psychology. Like Behe, he writes learned books asserting that ID is three things: a scientific research program investigating the effects of intelligent causes; an intellectual movement challenging Darwinian orthodoxy; and a way of understanding divine action. His writings have often been more obviously theological than those of others in the movement. Though he has worked with evangelical Protestants, his religious affiliation is the Eastern Orthodox Christian Church.

From 1999 to 2005, Dembski was on the faculty of Baylor University, where he attracted much attention and controversy. He was initially invited to Baylor by its president, Robert Sloan, a Baptist minister who wanted to reaffirm the institution's Christian heritage and further honor the Hungarian physical chemist and philosopher Michael Polanyi (1891–1976). Dembski had impressed Sloan's daughter when she met him at a Christian camp near Waco, Texas, and Sloan believed he would be an important asset to Baylor.

Following Sloan's directives, Dembski organized the Polanyi Center, which he identified as "the first Intelligent Design think tank at a research university." He hired a single colleague, Bruce L. Gordon, as assistant. Gordon, however, was not subjected to the usual university employment procedures of a search committee and departmental interviews, making his position vulnerable when the faculty started taking a close look at the center, which it had never endorsed. Dembski's operations soon became a subject of heated controversy on campus, with the faculty identifying his center as fringe thought, contrary to the university's stated mission. The faculty further reminded the administration that an interdisciplinary Institute for Faith and Learning already existed on campus, making the stated work of this new center redundant.

Events came to a head when Dembski in 2000 organized a conference called "Naturalism in Science," receiving a generous grant from the Templeton Foundation, a well-endowed foundation known for its support of scholarly religious endeavors. The conference also had the support of the Discovery Institute. As a whole, the Baylor faculty boycotted the conference, and a few days later the faculty senate appealed to President Sloan to dissolve the Polyani Center or merge it with the Institute for Faith and Learning. Very reluctantly, Sloan, who regarded the action by the faculty as an assault on academic freedom, agreed to a review that finally resulted in the merging of the center and the institute.

Dembski did not accept this change quietly. His heated exchanges with faculty members were highly publicized. Though he remained an associate research professor until 2005, he was given no courses to teach. Instead, during what he later called his "five-year sabbatical," he wrote books advancing ID and accepted speaking engagements around the country.

In 2005, Dembski was appointed Carl F. H. Henry professor of theology and science at the Southern Baptist Theological Seminary in Louisville, Kentucky. The seminary, which had already been purged of its more liberal faculty, was heavy with young earth creationists, whose views did not totally accord with Dembski's. He remained there only a year, before leaving to become professor of philosophy at Southwestern Baptist Theological Seminary in Fort Worth, Texas.

Throughout his many controversies, Dembski's chief disagreement with Darwinism is over specified complexity. He finds patterns in organic life that are both complex and specified, that he feels could not have been formed by random, mindless forces. Specified complexity demonstrates a clear purpose in its design. For example, a random set of letters might be formed by a chimpanzee using a word processor, while a poem, in which the letters are arranged in words that not only convey a clear meaning but subtle nuances as well, demonstrates specified complexity.[10] Behe's favorite examples of irreducible complexity—the human eye and the blood clotting mechanism—are also examples of specified complexity.

But Dembski's main value to the ID movement may be his role as a dramatic critic of the current scientific establishment. He does not feel that

Darwinism is sufficiently supported, at least in all its major parts, by the scientific evidence. He regards Darwinian evolution as a dangerous ideology promoted by a liberal elite who espouse atheism and are intent on forcing their will and beliefs on everyone. He also wishes to see ID active in a productive scientific research program. Though he does not believe this will happen soon, because the current generation of scientists is doctrinaire, his hope is that a new generation will arise more open to the mysteries of the universe.

MICHAEL DENTON (1943–)

Two of the most interesting figures associated at various times with intelligent design are brilliant iconoclasts who would not fit easily into any organized movement. They are Michael Denton and David Berlinski. It is Denton who has presented the most convincing argument for the "fine-tuning" of the universe and what it implies for biological evolution and the emergence of humanity on planet earth. A British-Australian biochemist with a PhD from King's College London (a college of the University of London), Denton challenged Darwinism and stated his evidence for supernatural design in his 1985 book *Evolution: A Theory in Crisis.* The book led many into the ID movement, and Denton was for a time identified with the Discovery Institute. Now his ideas have diverged too far for him to be comfortable in the institute.

Denton currently accepts evolution, but with a difference. He opposes the special creationists, which make up the majority in the ID movement, though he still believes that evolution is clearly directed and the origin of life must of necessity occur when the conditions are exactly right. In his more recent publications, he has accepted the adequacy of the Darwinian mechanism of evolution, though he still denies that randomness accounts for the biology of organisms. Instead he has proposed a "directed evolution," working out his newer theories in *Nature's Destiny* (1998). Life, he believes, did not exist until the initial conditions of the universe were fine-tuned; yet they were waiting for life to appear. He appeals to the special character of DNA evidence, observing that DNA sequencing demonstrates that the genomes of all organisms are clustered very close together in a tiny region of DNA sequence space. All life is fragile and requires precise environmental conditions for it to flourish.

Denton's writings demand serious attention. He has proven his independence as a thinker and his ability to modify his theories after further examination and research. Although he now receives more respectful attention from other scientists than do the fellows of the Discovery Institute, it is impossible to exaggerate the influence he has had on the institute.

DAVID BERLINSKI (1942–)

Possibly the most intriguing personality associated with the Discovery Institute is senior fellow David Berlinski. He is not a Christian and sometimes even identifies himself as "a Jewish agnostic." In an interview, he has admitted: "I have no religious convictions and no religious beliefs. What I do believe is that theology is no more an impossible achievement than mathematics."[11] Unlike the other fellows, Berlinski cannot be accused of trying to infiltrate the sciences with a particular religion.

Berlinski was born to Jewish German refugees from Nazi Germany and spoke German before any other language, even as he grew up in New York City. He is the sort of intellectual often referred to as a Renaissance man, who has excelled in a number of different disciplines and is always restlessly exploring more areas of learning. He currently writes detective stories, and his work has appeared in several well-received literary anthologies, while his scientific writing has been described as "peculiar, odd, or mischievous." He is known for his wit and his delight in puncturing the pretenses of pompous learned folk.

Like other Discovery Institute fellows, Berlinski has impeccable credentials: a PhD in philosophy from Princeton University and a postdoctoral fellowship in mathematics and molecular biology from Columbia University. His published professional work includes studies in system analysis, differential topology, theoretical biology, analytic philosophy, and the philosophy of mathematics. In his spare time, he has written three novels. His teaching career has taken him to Stanford University, Rutgers, The City University of New York, and the Université de Paris. As a maverick intellectual, he boasts that he has been fired from almost every teaching position he ever held. After working for a time in Austria, he settled in France, where he now lives and works.

In making common cause with the Discovery Institute, Berlinski shares the belief that the evidence for evolution is presently inadequate and the scientific establishment is rigidly dogmatic. Still, he often deviates from the views of his ID colleagues, describing his relationship with some of them as "warm but distant" and adding that he has the same public relationship with them that he has with his ex-wives. He is perhaps most valuable to them as a critic of all systems, a compulsive iconoclast. In an article for *Commentary* titled "The Deniable Darwin," he gave the following reasons for his skepticism:

1. The appearance "at once" of an astonishing number of novel biological structures in the Cambrian explosion.
2. The lack of major transitional fossils.
3. The lack of recent significant evolution in sharks.

4. The evolution of the eye, particularly the value of "part of an eye."
5. The failure of evolutionary biology to explain the range of phenomena from the sexual cannibalism of redback spiders to why women are not born with a tail.[12]

Berlinski's reviewers sometimes complain that his arguments are hardly new, but they cannot deny that his style is engaging and witty, a rarity in writings on evolution, either pro or con. He is unimpressed by the arguments of prominent atheistic scientists such as Daniel Dennett and Richard Dawkins. Dennett he dismisses with contempt or indifference, but Dawkins he finds fascinating in a perverse way. He describes Dawkins as "louche . . . fascinating and repellant . . . an intellectual fanatic." Dawkins, he contends, uses the same tired old arguments with which Bertrand Russell used to try to impress his mistresses!

Berlinski is an old-style combatant in debates. His disdain for Darwinists is evident, finding them filled with indignation and intent on promoting their theory because so much money, prestige, power, and influence are at stake. It is easy to get the impression that Berlinski dislikes evolutionists more than evolution itself. He alleges that there is nothing so fantastic that physicists and biologists cannot accept it, that they are the sort of people used-car salesmen love. And he reminds academicians who sneer at religious experience and moral absolutes that they need to remember who pays their salaries.[13]

SIR FRED HOYLE (1915–2001) AND CHANDRA WICKRAMASINGHE (1939–)

No discussion of critics of Darwinism would be complete without mention of Fred Hoyle and his student, Chandra Wickramasinghe, and their intriguing, unorthodox theory of how life originated on earth.

Hoyle was a leading astronomer of the 20th century, famous for his contribution to the theory of stellar nucleosynthesis. Most of his work was completed at the Cambridge Institute of Astronomy, which he directed for a number of years. He had a rich imagination and published several science fiction novels, some of them cowritten with his son Geoffrey. He is further remembered for jokingly coining the phrase "Big Bang" to describe the cosmological theory he personally rejected. The phrase is now in common use. Repeatedly, he demonstrated his independence as a thinker and his willingness to defy scientific orthodoxy.

Although he had been an atheist in youth, Hoyle's scientific work convinced him that a super-calculating intellect must be at work in the universe. This conclusion, he admitted, left him "greatly shaken." From that point on, he believed in a power of greater than human intelligence and developed, with his student Chandra Wickramasinghe, the theory of "panspermia," proposing that life evolved in space and spread to earth and throughout the universe by viruses arriving via comets. In his 1982 book *Evolution from Space,* he expanded his theory of an Intelligent Originator. Rejecting the Darwinian theory of directionless evolution, he made the striking analogy of a tornado sweeping through a junkyard and assembling a Boeing 747 from the scrap found there. He thought the random emergence of life, according to the Darwinians, equally unlikely. Though Hoyle would hardly have identified with the Discovery Institute, some of his ideas have been of enormous use to the fellows there.

Chandra Wickramasinghe, the most famous student of Fred Hoyle, was born in Sri Lanka but has made his career in England, where he is professor of applied mathematics and astronomy at Cardiff University and director of the Cardiff Centre for Astrobiology. It was his work with Hoyle on infrared spectra of interstellar grains that led to the development of their theory of panspermia.

Wickramasinghe is also a pioneer in the science of astrobiology. Like his mentor, he is a man of rich imagination. In addition to his many scientific papers and more than 25 books, he is a poet of note. During the scientific creationist trial in Arkansas, he was the single scientist testifying for the defense whose world renown was acknowledged by the scientific establishment. Though the press was fascinated with him and his challenge to the Darwinist monopoly in the classroom, his theory of panspermia seemed so strange to the Arkansas courtroom that his testimony had no effect on the outcome of the trial. Yet he remains one of the most intriguing and unusual figures in contemporary science.

NOTES

1. Daniel Radosh, *Rapture Ready!* (New York: Scribner's, 2008), 276–294.

2. For a witty overview of literature favored by the creationist-religious subculture, see Radosh, 88–117.

3. A fuller definition, widely distributed by the Discovery Institute states that "The scientific theory of Intelligent Design holds that certain features of the universe and of living things are best explained by an intelligent cause, not an undirected process such as natural selection. . . . Intelligent Design theory does not claim that science can determine the identity of the intelligent cause. Nor does it claim that the intelligent cause must be a 'divine being' or a 'higher power' or an 'all-powerful force.' All it proposes is that science can identify whether certain features of the natural world are the products of intelligence." Quoted by Hunter B. Rawlings, III, State of the University Address, Cornell University, October 21, 2005. Available at www.cornell.edu/president/announcement-2005-1021,cfm.

4. This definition of irreducible complexity originally appeared in Michael Behe's book *Darwin's Black Box* (New York: Free Press, 1996) but has been quoted in a variety of publications.

5. Dembski has used this example several times. See his Web site, www.designinference.com, for more information.

6. For more information about the creationist activities of Duane T. Gish, see Ronald L. Numbers, *The Creationists* (New York: Alfred A. Knopf, 1992), 224–334.

7. *Skeptic* 5(2) (1997), 37–41.

8. The Wedge Document is widely seen as the work chiefly of Professor Phillip Johnson and has been endorsed by the ID movement. In carefully constructed language, it states two goals for the movement. First, it seeks the defeat of scientific materialism to which it attributes a decline in moral, cultural, and political standards. Second, it wishes to replace materialistic explanations with a theistic understanding of nature and the human place in nature. Perhaps more than anything else, the Wedge Document has aroused the suspicions of those who accuse ID of being a religious movement disguised as scientific.

9. www.lehigh.edu/~inbios/news/evolution.htm.

10. For a fuller statement, see William Dembski, *Mere Creation* (Downers Grove, IL: InterVarsity Press, 1998).

11. Jonathan Witt, "An Interview with David Berlinski: Part One," www.edthefuture.com/2006/03, 3.

12. *Commentary* 101(6) (June 1996).

13. Witt, 2.

6

Responses to Intelligent Design

GENERAL CRITIQUE OF ID

Although Darwin's theory of evolution cased great controversy when first introduced, it is almost universally accepted today among scientists. Of course, new research has modified the original theories, just as it has given added weight to the basic concepts. Research predicated on evolution goes on constantly in science laboratories around the world. Although there are still admitted gaps in the fossil record, the doctrine of punctuated equilibria is hotly debated, and even social Darwinism and eugenics have been generally discredited, there appears to be no serious scientific dispute about the validity of evolution itself. Biologists and paleontologists point out that numerous gaps that existed in Darwin's time have gradually been closed, as new discoveries are made each year.

Even if masses of people still do not acknowledge their descent from lower species, they have not hesitated to avail themselves of the medicines that have developed through evolutionary biological research. While benefiting from evolutionary theory, it may seem ironic that such hostility still exists not only in the United States but seems to be growing among groups in Australia, Korea, Russia, and Turkey. The antievolution movement remains well funded and not without influence.

Although evolutionary theory does not appear to encourage any religious affirmation, it is frequently observed that many orthodox believers are evolutionary scientists: Jews, Muslims, Hindus, and Eastern Orthodox, Roman Catholic, and Protestant Christians are all represented in the major scientific communities. Europe seems to be much less concerned than the United States with the alleged conflicts between science and faith, only in part because Europeans are more secularized. Even European countries that are still religious, such as Poland, do not offer fierce resistance to evolution. And the Christian churches that do persist in Europe—some of them officially established as part of governments—are no longer at war with science. The Vatican itself has made its peace with evolution.

Though many scientists concede that their disciplines do not exclude religious belief, they are practically unanimous in the opinion that all religious concepts of the supernatural should be kept out of the science classroom and the research laboratory. First, the scientific method does not involve the supernatural. Second, there is not enough time allotted in the average science classroom for all possible theories of origin to be taught. In a pluralistic society, whose theories would receive attention? Would the cosmology of Native American tribes be appropriate, or that of Buddhists, Hindus, or Zoroastrians? Third, teachers of science are not trained or qualified to delve into issues of theology or philosophy. And, finally, has it not already been established in the courts that religious indoctrination does not belong in public school classrooms in our pluralistic society?

Among the objections to creationism and intelligent design in the public schools is the fact that the antievolution movement is a blanket covering a variety of points of view, from young earth creationists to agnostic scientists who have serious problems with evolution. It is impossible to determine precisely what the movement would insist upon including if allowed into science classrooms. ID literature, according to its opponents, includes outdated science, misinformation, along with misquotations and misrepresentations of the work of numerous evolutionary scientists. A few acknowledged scientific mistakes and hoaxes, such as that of the Piltdown Man, are presented as if they were characteristic of the work of evolutionary scientists.

No important scientific developments nor applications have as yet been founded on either creation science or intelligent design. Although there is some merit in the suggestion that it is desirable to "teach the controversy" in advanced or upper-division high school classes as an inducement to creative thinking, ID singles out for attention only one controversy, its own, ignoring others where there is serious debate among scientists. Finally, while ID is widely labeled bad science, a large group of contemporary theologians find it equally bad religion, relying heavily on emotional appeals and heated rhetoric rather than logical thinking or sound theology.

It would appear to many opponents of creationism and intelligent design that the battle should by now have been won, after so many victories in the courts. Although the 1925 Scopes trial in Dayton, Tennessee, was legally a victory for anti-Darwin forces, it did solidify resistance to the ban on evolutionary teaching in schools, and the Tennessee law was eventually overturned, however tardily. In Arkansas, Pennsylvania, Kansas, and Ohio, creationism and ID suffered devastating legal setbacks. And in 1987, the final victory seemed to have been won when the Supreme Court declared that laws mandating the teaching of creationism violated the First Amendment to the U.S. Constitution. Both scientific creationism and intelligent design in public schools should now be dead issues, according to the American Civil Liberties Union and its supporters. Though ID advocates had been a bit harder than young earth creationists to refute, and their demands in the classroom had been relatively modest, judges as well as scientists have been convinced that their goals are more religious than educational. Their theories are believed to present no serious challenges to legitimate science and have failed to meet the basic test of a genuine scientific theory; they can be neither tested nor falsified. All the major objections to evolution would seem to have been decisively answered.

ID speakers have frequently demanded of evolutionists the answers to several questions they consider prickly. Evolutionists have responded with their own set of questions. How do ID proponents explain the convincing laboratory experiments with fruit flies that have shown evolutionary adaptations to differing temperatures, or with honeybees that have modified to become more competitive, or with guppies that have experienced color modification to elude new predators? How can the emergence of peppered moths in response to environmental changes brought on by the Industrial Revolution be explained? How have these creatures developed their more effective camouflage, changing their color as trees in England darkened from industrial pollution? If evolution does not occur, how can mutating viruses be explained?[1]

Questions about the Designer of all this also demand to be answered. Precisely what was the Designer's method? Why did the Designer leave so many imperfections in creation, such cruelty? Is the Designer a cosmic sadist? Why did the Designer make so many unnecessary organs in beasts and humans? Were all species made at once, or over a long period of time? Was the first human a child or an adult? And then there is that old, rather silly question about whether or not Adam had a belly button!

Other objections are voiced by ID opponents. Because virtually all of modern science, with its immense advances, operates through some understanding of the forces of evolution, to deny Americans the fullest, most modern science education is to dilute necessary human knowledge, with dire results in the future, as the United States falls behind other nations in scientific developments. If ID succeeds in establishing itself in high schools, students will be ill prepared to pursue science majors in college, and if colleges themselves are infiltrated by ID, major careers in science will become impossible for their graduates.

Possibly more serious still are the accusations of deceit and dishonesty leveled against creationists and ID by their adversaries. In heated discussions of issues where contending parties have much at stake, it is not unusual for arguments to become acrimonious, and the ID-evolution debates are no exceptions. They are filled with ad hominem attacks. Ashley Montagu, writing in 1984, said that creationists might call themselves scientists and refer to their "manipulations" as "creation science," but they are no more scientific than Christian Scientists or Scientologists.[2] He found their teachings to be religious, and concluded that they were dishonestly trying to evade the First Amendment to the Constitution by claiming scientific foundations for them.

While Montagu felt there need be no real conflict between evolution and religion, he did find scientific creationism incompatible with science, enlightened religion, and civility. He believed that fundamentalism, a peripheral sect of Christianity, was seeking to impose its particular creation myth on the public as an alternative to scientific explanation. In their insistence of having their myth alone taught in the public schools, alternate religious as well as scientific points of view were being excluded.

Kenneth R. Miller has not been as ferocious in his attack on ID as was Montagu, and he usually moderates his critique with humor. He feels that one reason for ID's substantial success—and he is forced to admit that various creationists have been very successful in reaching the general public—is that many scientists have been unable to adequately present their findings to the average American audience. Admittedly, scientific education has been inadequate, and an otherwise informed and intelligent public needs scientific speakers and writers who can communicate effectively with a lay audience. Even though creationists have failed to rationally defend their views, the liveliness and combativeness of their speeches and publications have made them enormously entertaining. At the same time creationists of all persuasions have failed to acknowledge the vast amount of research that supports evolution, they have been unable to provide an alternative theory of natural history that fits known facts. They simply deny scientific findings that do not accord with their preconceptions, including current explanations of radioactive decay, the constancy of the speed of light, and equilibrium thermodynamics.[3]

The late Isaac Asimov, ever-prolific writer on any subject that held his attention, once satirized creationists in his science fiction magazine, facetiously proposing a new genre of "creation science fiction." The arguments of creationists carried no weight with him. He felt they used fallacious techniques of argumentation: from analogy, from general consent, from questionable authority, from the imperfections of opposing arguments, and by belittlement and irrelevance.[4]

AN ATHEISTIC ATTACK ON CREATIONISM AND ID

One of the most aggressive critics of all forms of creationism, especially ID, is the English atheist, media celebrity, and entertaining popular writer Richard Dawkins. A rarity, the scientist who can communicate complicated information to a general audience with a certain amount of humor, he is also an aggressive atheist on a mission to enlighten theistic readers. He considers those who oppose Darwinism benighted at best and often consciously dishonest. Not only does he attack creationists, but he rarely spares other scientists with whom he disagrees on any point. In his books, he argues that the scientific explanation is now totally adequate to explain natural phenomena, without postulating any supernatural creative intelligence. In the preface to his best-selling book, *The Blind Watchmaker* (1986), he writes:

This book is written in the conviction that our own existence once presented the greatest of mysteries, but that it is a mystery no longer because it is solved. Darwin and Wallace solved it, though we shall continue to add footnotes to their solution for a while yet. I wrote the book because I was surprised that so many people seemed not only unaware of the elegant and beautiful solution to this deepest of problems but, incredibly, in many cases actually unaware that there is a solution in the first place.[5]

It is not surprising that strong spokespeople for controversial opinions raise heated responses. Some opponents of creationism and ID fear that Dawkins, by his very eloquence and celebrity, attracts undue attention and does his causes more harm than good. Christians are understandably enflamed by his attacks. Alister McGrath and Joanna Collicutt McGrath, in their 2007 book *The Dawkins Delusion* (which was occasioned by another best-seller by Dawkins, *The God Delusion*), reject the contention that religion is a delusion perpetuated on infantile, irrational people. The McGraths, who are British like Dawkins, are students of chemistry, molecular biology, and the psychology of religion. They take Dawkins to task for his one-sided attacks on religion, his limited concept of the God he denies, and his simplistic contention that religion is "a virus of the mind." While Dawkins believes that religion almost invariable leads to violence, citing such historic events as the Crusades and the Inquisition, he oversimplifies motives and events, failing to acknowledge the many social advances and instances of kindness that have resulted from religious sentiment. When he reads the Bible, he does recognize that it contains much splendid poetry and even concedes that, because of its literary value, it should be a part of the education of schoolchildren. Still, he focuses on passages he finds shocking to modern sensibilities—such as the harshness of parts of the Mosaic law and the brutality of the Israelite conquest of Canaan—to the neglect of reassuring and elevating messages of scripture.

The McGraths conclude their response to Dawkins's books by observing:

Many have been disturbed by Dawkins's crude stereotypes, vastly oversimplified binary oppositions (science is good; religion is bad), straw men and hostility toward religion. Might *The God Delusion* actually backfire and end up persuading people that atheism is just as intolerant, doctrinaire and disagreeable as the worst that religion can offer?[6]

NOTES

1. For a thorough explanation of these changes, see Jerry A. Cone, *Why Evolution Is True* (New York: Viking, 2009).

2. Ashley Montagu, ed., *Science and Creationism* (New York: Oxford University Press, 1984), 7.

3. Kenneth R. Miller has provided a humorous response to the favorite ID comparison of "irreducibly complex" organs such as the eye to a mousetrap. Eliminate one part and the whole becomes useless, according to a favorite argument of Michael Behe. Miller has suggested that parts of the mousetrap could be dismantled to make a fine spit-ball launcher for dull days in school study hall. If the spring is detached from the clip, a working key chain remains. With the attachment of a magnet, it becomes a refrigerator clip. Remove the hold-down bar and one has a toothpick, while three parts of the trap alone can function as a tie clip. Miller, *Only a Theory: Evolution and the Battle for America's Soul* (New York: Viking, 2008), 53–57.

4. Isaac Asimov, "The 'Threat' of Creationism," in *Science and Creation,* ed. Ashley Montagu (New York: Oxford University Press, 1984), 183–190.

5. Richard Dawkins, *The Blind Watchmaker* (New York: W.W. Norton, 1986), XI.

6. Alister McGrath and Joanna Collicutt McGrath, *The Dawkins Delusion* (Downers Grove, IL: InterVarsity, 2007), 97.

7

Religion and the Courts

In the first decade of the 21st century, members of the Roman Catholic Church dominate the U.S. Supreme Court. There are two Jewish justices, one Protestant, and the rest, including the Chief Justice, are Catholic. Interestingly, this is a matter of no concern to evangelical Protestants, who now tend to agree with the official Catholic position on many of the social issues dividing American society. But the composition of the Court is significant. Until the 20th century, the United States was regarded as a Protestant country. The founding fathers came out of Protestant Christian traditions, although many of them were influenced by the rationalism of the 18th century and its Deism. Still, until the first decades of the 20th century, Protestant Christianity was a recognized part of American life. Ministers offered prayers at public functions; school textbooks reinforced values identified with Protestantism; and the King James version of the Bible was regularly read at the beginning of each school day throughout the country.

THE BACKGROUND

The Establishment Clause of the First Amendment to the Constitution was framed with the Anglican type of church in mind. An established church, such as England maintained, with the ruler at its head, was anathema to the founders. Even more anathema was an Italian pope in a city like Rome. Thomas Jefferson expressed equal disdain for kings and bishops. The First Amendment was not intended, according to historians, to cripple religion or even eliminate religious establishments that were already present in certain states. The purpose was to avoid a national church, leaving individual states free to make their own determinations. The free exercise clause prevented government persecution of dissenting groups, which had been a frequent occurrence in England. The degree to which it also released citizens from federal laws that violated their consciences was yet to be determined by the courts. In the early years of the nation's history, some state laws still placed obstacles on atheists and people who were not Protestant, since the foundation of American culture and social life was recognized to be Protestant.

Early in the 19th century, state religious establishments disappeared, though a total secularization of society was still not the general intention. The de facto Protestant establishment still existed, and most educational institutions, as they came into being throughout the century, retained a distinct Protestant flavor. Prayers and Bible readings were led each day by teachers or students. Prayers were offered at legislative assemblies, and many had their own chaplains. Thanksgiving, a uniquely American religious commemoration, along with Christmas and Easter were official holidays in which schools dismissed and most people got leave from work. Laws against blasphemy were taken seriously. In numerous places, limitations on alcoholic beverages were later enforced, despite the fact that wine was part of the central Christian sacrament. Sabbath-keeping was widely honored. There were unwritten religious qualifications for public office.

Throughout the country, Mormon polygamy was opposed with horror and titillation, based not only on the Christian preference for monogamy but also on the perception that Mormonism was outside the pale of Protestant Christianity. A quasi-pornographic literature appeared, allegedly detailing adventures in Mormon harems in Utah or nefarious doings in Catholic convents and monasteries, especially in the remote Canadian province of Quebec.

During the de facto Protestant period, a generalized Christianity, only vaguely sectarian, emerged for public consumption. Because the King James version of the Bible was widely revered, and admittedly had exerted a tremendous influence on English language and literature, it was regarded as nonsectarian. Most people did not realize, until society became more diverse, that the King James version was one that in some ways offended Roman Catholics. Jewish students, of course, had their own translations of the Bible and accepted only what Christians referred to as the Old Testament. Prayers in Jesus' name, customary with all Christians, became a problem when Jewish students entered the classrooms.

Early in the 20th century, new forces were already challenging the Protestant establishment. Darwinism and the growth of theoretic and applied science were only one reason for the decline of religion in the public arena. There was immense growth in U.S. education. Universities, most of which had been founded by religious bodies to train their ministries, were moving away from religious control and emphasis. Pragmatic social rules were now enforced rather than those dictated by religious imperatives. As the law itself became more secularized, intellectual criticisms of religion and reactionary practices associated with religion were voiced, especially in colleges and universities.

The influx of immigrants from Roman Catholic countries changed the nature of Christianity in the United States. Persecutions in Europe also meant that many Jews found refuge in the country. Jewish and Roman Catholic immigrants were anxious for their children to receive the kind of education that would enable them to succeed in the new land. But both groups were also deeply concerned that the Protestant influence in the public schools would alienate these children from their own heritage. Many immigrants came from lands where education had been in the hands of the clergy, and they associated learning with the church. Fearing the contamination of their faith, Roman Catholics developed the largest system of parochial schools in the country. Lutherans, never considering themselves exactly Protestant even though Martin Luther had started it all, soon established their own schools, as did the orthodox Jews. It was not long before these groups were agitating for some public aid to their school systems. Numerous cases reached the courts.

Opponents of state support for parochial schools have been fond of quoting Thomas Jefferson's famous letter to Danbury Baptists, in which he used the phrase "wall of separation between Church and State." Many people inaccurately believe the phrase is actually in the Constitution, and court decisions have reinforced this misunderstanding. Courts have generally allowed aid to parochial schools only if a secular purpose for such aid could be clearly established. For example, fire and police protections have been generally permitted. Courts have sometimes allowed states and communities to provide transportation for students to parochial schools and have approved subsidies for nonreligious textbooks. Bible classes offered in public schools on a voluntary basis by ministers of different denominations have not fared so well in the courts. Released time for such classes during the regular school day has also been problematic, as have been religious organizations holding meetings on school property. One court allowed students to organize religious associations that met after school on school property, as long as faculty members did not initiate these programs. Many people, especially those who had experienced Bible reading and prayer in their own elementary and secondary schools, concluded that U.S. courts were now hostile to religion.

Education has not been the only area in which U.S. law has found itself in conflict with religious sentiment. A few religious groups have not only tolerated but advocate polygamy. In examining such cases, courts have determined that, while government must not attempt to regulate religious belief, it still has some control over religiously motivated actions that may violate the ethical standards of the dominant population.[1] Court decrees have been more favorable to Seventh-Day Adventists and Jews who have refused to work on the Sabbath, finding insufficient reason for employers to refuse to accommodate these employees. When Amish parents refused to send their children to school past the eighth grade, the courts sided with them, determining no overriding reason to enforce further education. Yet when the Amish argued that the paying of social security taxes was contrary to their belief, the courts decided against them.[2]

But court cases involving the education of young children have been the most lastingly significant. In *Lemon v. Kurtzman* (1970), the court set a precedent. A three-pronged test for possible government involvement in parochial school programs resulted. To qualify for public aid, the court decreed that a program must (1) have a secular purpose that neither endorses nor deters religion; (2) must have an effect of neither endorsing nor inhibiting religion; and (3) must avoid creating any entanglement of government with religion.[3]

Religion must also, according to the courts, be dealt with very carefully in public schools. Eventually the courts defined a policy by which religion, if included in any public school's curriculum, had to be impartially incorporated into a secular program of education.

THE FAMOUS COURT CASES

Creationism and intelligent design in the public schools have been tested in a number of highly publicized court cases that have attracted journalistic coverage all over the world. The most famous remains the Scopes trial in Dayton, Tennessee, in 1925, celebrated in story and song, and popularly known as "the Monkey Trial." The case widely referred to as "Scopes II" took place more than fifty years later in Little Rock, Arkansas, and the third most publicized case unfolded in Dover, Pennsylvania, in 2005.

Though initially Americans were less preoccupied with Darwinism than with social issues that impinged upon religion, by the 1920s, in part because of its presumed or real association with unfriendly Germanic ideas, Darwinism had become a major concern. The public school system was a chief battleground. People were now attending high schools in much greater numbers than ever before, with the public schools playing an increasingly influential role in U.S. society. As these schools proved skillful at making Americans out of people from many lands and traditions, distinct American myths were perpetuated: George Washington and the cherry tree; Honest Abe reading by firelight and splitting rails; presidents born in log cabins.

Children ate cherry pie on Washington's birthday and staged pageants on Flag Day. With religious affiliations and ethnic customs more diverse than ever before, the schools became the real melting pots of the society, where a common Americanism was celebrated.

As more people attended high schools and were exposed to advanced learning, the teaching of evolution became an issue in these cherished public schools. This was particularly pronounced in communities of the American South and the heartland, where the population was more homogeneous and the churches were the most influential institutions in the community.

Europeans had long labeled Americans as the least philosophically sophisticated people in the developed world, and evolution was associated with ideas of class and privilege. The timing of the introduction of evolution into the schools was also inauspicious. Militaristic Germany was perceived throughout the first half of the century as the enemy, and the United States would fight two world wars to rid the world of German domination. Germany was also the source of the higher criticism of the Bible that was beginning to disrupt and divide churches. Americans struggled with conflicts over religious liberalism, which accepted many of these new German ideas, and was regarded as a more pernicious influence even than atheism, because it seemed to create a "fifth column" within the churches themselves. Evolution, which had been used to justify German militarism, was equally subversive to a nation that wanted to hang the German Kaiser.

In 1922, the Kentucky legislature only narrowly defeated the first state antievolution bill. The movement grew, nevertheless, especially after William Jennings Bryan, one of the best-known politicians in the country, lent his support and his powerful oratorical skills to it. During this period, the Florida legislature officially condemned the teaching of evolution in the public schools, while Oklahoma prohibited the mention of evolution in textbooks used in the state's public schools. However, that law was repealed in 1926 and was never without forceful opposition. An antievolution bill was discussed and widely supported in West Virginia but failed to pass the legislature. Three states—Tennessee, Mississippi, and Arkansas—did pass laws in the 1920s banning the teaching of evolution in classrooms supported by taxpayers.

But it was Tennessee that gained worldwide attention with its antievolution law, introduced as the Butler Bill. The Tennessee legislature passed the law, initiated by John Washington Butler, a farmer from east Tennessee who had just been elected to the state legislature, in January 1925. Though there was isolated dissent, from places like Vanderbilt University in Nashville, with the support of many voters, the bill had hurriedly passed, with a vote of 71 to 5 in its favor. It stipulated that theories of evolution that contradicted the biblical account of creation were illegal in any publicly supported schools, including state universities. Governor Austin Peay signed the bill, which many people, probably including Governor Peay himself, regarded as only a symbolic gesture to pacify a large, deeply committed segment of the population. Though Peay was a devout Baptist, he does not appear to have expected or desired that the law be enforced. The bill made the teaching of evolution only a minor offense, which the state colleges and universities would probably ignore.

Social and Religious Background of the Scopes Trial

The American Civil Liberties Union (ACLU) had been looking for a test case, hoping to discredit all antievolutionary laws with an eventual appeal to the United States Supreme Court. The ACLU, first known as the National Civil Liberties Bureau, was an emerging organization, founded in 1917, first to protect conscientious objectors to World War I. A few years later, it would give its attention to religious incursions into public institutions, eventually waging war against Bible readings, prayer, denominationally neutral Bible classes, and even the singing of Christmas carols in public schools.

The ACLU in the East, where its strength lay, took note immediately of the new Tennessee law and placed advertisements in the Chattanooga, Tennessee, newspapers and elsewhere offering support for a test case. At the same time, Dayton, a small town nestled in the Tennessee mountains, was seeking attention. Unlike many remote, provincial communities of the Cumberland region, Dayton was populated by productive, progressive citizens. But the town had experienced a recent economic decline. Several of the leading citizens, sitting around a table in the local drug store, designed a plan to put their town on the map by testing the law.

John Scopes, a local high school civics teacher and coach, not a native of the region but a recent graduate of the University of Kentucky, was approached. Although he had only substituted a few days for the ailing high school biology teacher and did not really remember if he had mentioned Darwin during that time, Scopes regarded the new antievolution law as a travesty and was pleased to cooperate. After all, his job was secure, with his principal and school superintendent in on the adventure, and if he were found guilty, which he surely would be, the maximum fine was only $500. Although that was a considerable sum in the mid-1920s, especially for a young teacher, it would be paid by his sponsors. Anyhow, he was a bachelor and did not intend to remain long in Dayton. He expected publicity, which was, after all, the point, but Scopes did not realize that, until the end of his life, he would retain a notoriety. Neither did the city fathers envision the dubious reputation their action would forever give their town.

William Jennings Bryan had been busy awakening the public to the alleged dangers of Darwinism, and he had lectured in Tennessee just before the enactment of the Butler Bill. His objections to Darwinism were, at least initially, more social than religious. He abhorred its uses by German militarists and saw in Darwinism a major disaster hanging over the human race. Although Bryan had trained as a lawyer, he had not practiced in many years and was, therefore, not really prepared for a keen test of wits in the courtroom. Still, he was a major celebrity, widely regarded as the most powerful orator in the country. As a young man, after delivering a startling speech at the Democratic convention, he had been chosen by an otherwise deadlocked convention as the party's standard bearer. When he lost that election, so great was his popularity that he had been chosen twice more as the Democratic Party's candidate for president. Defeated in his third campaign, he had served briefly as Woodrow Wilson's secretary of state and remained one of the most admired Americans. His social views were liberal, and it was partly through his influence that, in 1920, women had received the right to vote with the Nineteenth Amendment to the U.S. Constitution. Though he had no formal theological training, Bryan was a Presbyterian layman, who frequently lectured on religious issues. Not a fundamentalist in every respect, he still maintained a profound respect for the entire integrity of the Bible. Widely known as "the Great Commoner," he never wavered in his respect for the ordinary citizen, and he felt that citizens had a right to determine what their children were taught in the schools they supported. Bryan relished the opportunity to defend these beliefs in a highly publicized trial.

Clarence Darrow, though an equally impressive personality, lacked the common touch of Bryan, and it is easy to understand why the American Civil Liberties Union was reluctant to include him in their project. Known as a supporter of unpopular causes, he had in speeches already equated evolution, which he defended, with atheism. He had earned wealth and much publicity, not all of it favorable, during his legal career. In an earlier highly publicized case, he had defended Leopold and Loeb, two wealthy Chicago youth who had killed another young boy merely for the experience. Though they were convicted, Darrow's skill had saved them from the electric chair. An early political supporter of Bryan, Darrow had recently become disenchanted. Now he was eager to challenge Bryan and, at the same time, demonstrate publicly that fundamentalism was as ridiculous as he believed it to be. He also savored the publicity that he knew the trial would bring. Darrow was not always judicious in his pronouncements, but he was John

Scopes's own choice for his defense. In a long career, this was the only case Darrow ever took pro bono, hoping to go all the way to the U.S. Supreme Court with it.

According to Professor Edward J. Larson, who has written the most balanced and authoritative book on the trial, several factors came together to make the event so crucial that its effects would linger on for decades and it would become an American legend. Protestant fundamentalism was at the peak of its influence during the 1920s. In the major denominations (which would come to be designated "mainline"), it was seen as a counterweight to the religious liberalism that was creeping in, particularly from clergy who had been educated in major seminaries and were increasingly removed from the piety of their congregations. Developments in genetics had also made Darwinism essential in scientific studies, while a few decades before it had been scientifically controversial. It was now almost universally accepted in the world of scientific research. Also, until the 1920s, relatively few Americans attended high school. Now high school education was for the first time reaching the vast American public, where previously elementary education had been considered adequate, at least in many locales, the South being one of them. More and more children were being exposed to ideas their parents did not understand. As if this were not enough, the Roaring Twenties was a period of social disruption and tension. Puritanism was still strong in the South, where people shuddered at tales of the manners and mores of big-city life and demanded a return to normalcy, without any clear idea of what this was.[4]

The attack on Darwinian ideas was strong throughout the decade. Evangelists discovered it was a subject that aroused the faithful and not merely in the South. Among the leaders in the anti-Darwin movement were William Bell Riley of Minneapolis (the Northern Bible Belt), John Roach Straton from New York, and J. Frank Norris of Dallas. Riley, Straton, and Norris were highly influential Baptist ministers, though the conflict was by no means confined to that denomination. Institutions were also founded in good part to combat Darwinism. These included the World's Christian Fundamentals Association, the Bible Institute of Los Angeles, and the Moody Bible Institute of Chicago. The established denominations were bitterly split over evolution, which was a major subject for discussion in colleges with religious affiliation. Several conservative religious denominations established or enhanced already existing colleges of their own, where "the truth" would be taught in science as well as in Bible and literature classes.

More attention was given in these schools than ever before to literal interpretations of sacred scripture. Allegorical or symbolic approaches to various literary genres of the Bible were now out of favor. Receiving particular attention was the Genesis account of the seven days of Creation. Though some highly religious people, including Bryan himself, might accept the day-age theory or remind themselves that one day is as a thousand years in the eyes of the Lord, many were comforted by the belief that the literal Bible was as reliable a text on history and geology as it was a guide to ethics and their relationship with God.

As they defended their faith against increasing criticism, conservative Christians feared that if one part of the Bible were conceded to be fallible, it would all fall. The same Holy Book that described the creation of Adam in God's image also narrated the life of Christ and affirmed his divinity. Because the sole authority for evangelical Protestants was the Bible, any challenge to its authenticity cut at the foundation of their faith. Roman Catholicism and other high church bodies had less difficulty with Darwinism because they rested their faith not on scripture alone but also on holy traditions, decrees of church councils, and episcopal pronouncements.

Not merely the Southern and Midwestern Bible Belts, but the entire religious segment of the country was ablaze with the conflict. The Divinity School of the University of Chicago, distinguished educationally, was noted as a hotbed of liberalism, as was the New York bailiwick of the Reverend Harry Emerson Fosdick, an influential and highly opinionated preacher. Eventually the institutional leaders of most of the established Northern denominations sided with the liberals, though this was not true of large numbers of the faithful within their churches. Some people remained in conservative parties within their churches, a few entire denominations split over evolution and related issues, while in other instances, the faithful simply left for more congenial churches whose doctrines they could espouse.

While leading scientists might maintain a polite conversation with the liberals who attempted to reconcile Christianity with Darwinism, they regarded fundamentalists as a threat to the integrity of their disciplines. Conservative religion was viewed as a force impeding scientific education, if not necessarily scientific experiments themselves, in the United States. Numerous political figures, such as President Herbert Hoover, a Quaker, sided with the liberals. Later, when Hoover's presidency (1928–1932), however simplistically, became associated with economic disaster and the many "Hoovervilles" sprouted throughout the country during the Great Depression, fuel was added to the fundamentalist fire.

The Trial

Against a sometimes turbulent social background, the Dayton trial got underway. As the celebrity lawyers, Bryan and Darrow, arrived, a circus atmosphere prevailed in the town, which was flooded with press from all over the country and numerous foreign lands. Even the popular arts got into the act.

Charles Wolfe, a professor at Middle Tennessee State University in Murfreesboro, Tennessee, and a world-recognized authority on country music, has examined the influence of the Scopes trial on that emerging industry. By 1925, as Wolfe has observed, the recording industry was acknowledging the popularity of hill country music heard on Southern radio and at country hoedowns. The trial was timely, and the sentiments of country music artists, as well as their fans, were clearly with an inerrant Bible and a creation timeline of six days. While the world press was intrigued by the trial and the rustic atmosphere of Dayton, country artists, not only recognized a market but were committed as well to the cause. They issued a number of songs supporting the antievolutionist crusade. These artists knew that rural and small-town audiences, who loved their music, resented the perceived assaults on the integrity of the Bibles that graced their parlors. They also clearly perceived an insult to their region and way of life by the elitists from the North.

One of the popular songs about evolution was recorded by Vernon Dalhart (Columbia, CO15037-O) and composed by Carson Robison. Entitled "The John T. Scopes Trial," its tone and sentiment were clear from the opening lines:

All the folks in Tennessee are as faithful as can be,
And they know the Bible teaches what is right.

One of the first superstars of country music was Uncle Dave Macon, a favorite of the Grand Ole Opry in Nashville for many years. Always accessible to his fans, Uncle Dave held court to all who passed from the front porch of his home outside Woodbury, Tennessee, a few miles from Nashville. He was Bible-bred, occasionally preached a sermon, and was father to a family of preachers. For his first recording after the Scopes trial, he chose his original composition, "The Bible's True":

Chorus: I'm no evolutionist that wants the world to see,
There can't no man from anywhere, boys, make a monkey
Out of me.[5]

Uncle Dave's song, with all its non sequiturs and its rural dialect, was popular because of the charm and back-country appeal of its composer-performer. It did not mention the Scopes trial in particular, and for that reason its popularity continued even after the excitement of the trial died down. Because the audience for the Grand Ole Opry was primarily composed of rural and small-town folk who believed the Bible, relied on its wisdom, and were offended by what they understood of Darwinism, the Opry composers and performers continued to celebrate the Bible in their songs. For many years, it was difficult to distinguish between the specifically devotional sets on the Opry and those that specialized in songs about cheating and love sickness or wept over "that silver-haired Daddy of mine." For decades, the Opry remained a stronghold of conservative piety.

Back in Dayton, evangelists set up booths on the streets, peddling their books and pamphlets. Prayer services were held all over town, while visitors from outside the region savored the renowned Southern cuisine. A live monkey was paraded about the streets. The acerbic journalist from Baltimore, H. L. Mencken, arrived to write his diatribes against what he referred to as "the Bible and Syphilis Belt."

While the circus continued outside the courtroom, the trial itself lasted eight days. Scopes never took the stand, spending part of the time in the local swimming hole with the son of William Jennings Bryan, trying to escape the oppressive summer heat that descended on Dayton both figuratively and literally. The defense brought in noted scientists from around the country as expert witnesses, traveling at their own expense. Their testimony, which the judge declared irrelevant, went unheard, to the disappointment of many of the locals who had looked forward to learning something about Darwinism. The judge refused to declare Chapter 27 of the Acts of 1925 of the State of Tennessee, popularly known as "the Butler Bill" unconstitutional at the beginning of the trial, as the defense requested. To the further annoyance of the defense, the judge insisted the court be opened each day with a prayer, as was his custom, alternating among local clergymen. Scopes did not contest the accusations against him, and he is believed even to have talked some of his reluctant students into testifying that he had, indeed, taught Darwinism in the classroom. The dramatic high point of the trial, after the judge had recessed the jury, occurred when Bryan allowed himself to be put on the stand to be cross-examined by Darrow. Darrow succeeded in revealing not only Bryan's ignorance of modern science but the fact that he did not even know his Bible as well as had been assumed. Darrow was satisfied that he had, at least, managed to humiliate Bryan, though it is unclear that Bryan himself felt this to be true. Certainly his admirers did not agree, and, after the trial, there were many invitations for him to speak all around the area. Funds were also successfully raised to establish a Bryan College in Dayton, an institution that still exists.

Possibly because of the excessive heat and other exertions, on top of declining general health, Bryan died in Dayton a few days after the trial ended. His followers declared him a martyr to the cause, while his critics believed his death the result of his comeuppance at the hands of Darrow, with perhaps some overindulgence in hefty Southern food a contributing factor. Darrow returned to Chicago, where he would later provide hospitality to Scopes, after the young teacher left Dayton to enroll in the University of Chicago.

Scopes was, of course, found guilty by the jury, which could offer no other verdict, since even the defense conceded that Scopes had violated the law. The judge imposed the minimum fine of $100. This was later overturned by the Appeals Court on the technicality that the law required that a fine be determined by a jury rather than a judge. This deprived the ACLU of its opportunity to make further appeals and also saved the state of Tennessee additional embarrassment. The law did remain on the books, often ignored and never again enforced, until the 1960s.

Scopes did not lose his job but decided to accept a scholarship at the University of Chicago, training for a career as a geologist. Later, he became a successful petroleum engineer in Venezuela. In his later memoir, he admitted that he chose work in South America to escape further public attention. During the latter years of his career, he was an oil refinery manager in Louisiana. Although he had been reared by a Presbyterian mother and a free-thinking, British-born father, he converted to Roman Catholicism to please the woman he married, even while he acknowledged that he was never more than a nominal Catholic.

The result of the trial was that both sides declared vindication, if not full victory. But ultimately the teaching of science was the real victim. Dayton, unfairly, became a symbol around the world of obscurantism and ignorance. Books and plays were written that portrayed the city fathers of Dayton, or their fictional proxies, as the equivalent of the enemies of Galileo, conducting a modern Inquisition.

In later years, when Scopes was prevailed upon to dictate his memoirs to James Presley (the book was published in 1967), he described growing up in Paducah, Kentucky, and credited his father, a labor activist from England, and Clarence Darrow as the people who had influenced him most. After graduating from the University of Kentucky in Lexington, where he had completed a general education program, his plan had been to study law. In the meantime, with the necessity of earning a living, he had accepted the teaching position in Dayton, expecting it to last only a year or two. Because he did not regard teaching as his career goal, and because he was still unencumbered by a family, it had been easy for him to agree to test the Tennessee antievolution law.

In later decades, Scopes remembered the people of Dayton kindly. He had savored the natural beauty of the Tennessee mountain community, and it had offered some social opportunities for a young man. He remembered dating high school students only a few years younger than himself. (Apparently, that was more acceptable in those days than it later became.) He still could not remember ever mentioning evolution in the classroom and was not sure he had done so. The textbook from which he had taught, the state-approved *A Civic Biology*, admittedly contained a brief section on evolution. Widely used throughout the United States, the text also promoted eugenics and exalted the superiority of the Caucasian race, but even the ACLU had ignored that!

Though embarrassed by the excessive publicity he had received, Scopes did reap some benefits from his celebrity. Scholarships came his way, along with the lasting friendship of Clarence Darrow. Some years later, when *Inherit the Wind,* the play and motion pictures, appeared, he did take note. He even agreed to help with some of the publicity for the first major film. Names of people and places had been changed, but there was no effort to deny that Dayton and Scopes had inspired these dramatic representations. Although he was not angered by either the play or films, Scopes did point out inaccuracies in what the public came to know, or think it knew, about the trial. First, he did not consider William Jennings Bryan the fool or bigot that the play and film made him out to be. He conceded that his own rendezvous with Bryan had taken place at the end of the politician's long, productive career, when the man was no longer at the peak of his powers. Second, the film falsely depicted the people of Dayton as his foes, who jailed him and fired him from his teaching position. None of this ever happened. He was, instead, asked to return to his post for the next year, though he had declined in order to pursue graduate studies.

The major distortion that Scopes noted was in the depiction of his love life. In the dramatic adaptations, the hero is engaged to the daughter of a local minister. At the time of the trial, Scopes had no steady girlfriend, though a young woman he had dated only once did waylay him with a kiss for the benefit of visiting photo-journalists lurking nearby.

The play is still produced in regional and university theaters. It appeared during the McCarthy era and, according to its producers, was intended only as a parable, the real target being McCarthyism and the anti-Communist crusade then taking place. The best-known film production featured Spencer Tracy in the Darrow role and Frederick March as Bryan. Both were talented performers, leading men who had become strong character actors. Tracy came off as much more likeable than the real Darrow, while March's performance caricatured Bryan. The film slandered Southern people and the citizens of Dayton (though fictional names were used), who in real life were by no means the narrow-minded yokels represented.

Perhaps Scopes's final judgment on the people, place, event, and legend was expressed in the following words from his memoir:

Intolerance apparently plays no favorites. I have often said that there is more intolerance in higher education than in all the mountains of Tennessee. There is a tendency for educated people to insist that others less schooled should think as they themselves think. I wouldn't let anybody, whether he was from the Tennessee hills or the Harvard graduate school, control my thinking. By the same token, the Tennessee hillbilly and the Harvard professor have the same rights to their viewpoints, whether theirs coincide with mine or not.[6]

The Little Rock Trial

Another highly publicized court case, widely referred to as Scopes II, took place in 1982 in Little Rock, Arkansas. *McLean v. Arkansas Board of Education* involved the state's right to mandate that scientific creationism be taught alongside evolution. The law, Arkansas Act 590, appealed to an American sense of fairness and was based on the principle that students should be able to make up their own minds about opposing theories. This time the court ruled, in a precedent-setting decision, that the balanced presentation would constitute the establishment of a religious view because creationism necessarily involved a Creator, understood, at least in the Western world, to be the Judeo-Christian God.

The trial remains significant because it aired several issues. The act had not mandated the teaching of the biblical account of Creation, but actually disallowed it. It had further stipulated that evolution be taught, but alongside creationism. It did not refer to a deity or religious concepts. It did not even force teachers who opposed creationism to teach it, but permitted them to bring in a visiting teacher for that unit of instruction. Its supporters strongly denied that it was a fundamentalist or born-again Christian act. Fundamentalist Protestants, at least in earlier times, had been totally opposed to any teaching of evolution and wished to allow only the Genesis account in the classroom. Furthermore, one of the strongest promoters of the Arkansas law was Paul Ellwanger, a Roman Catholic layman rather than a Protestant of any persuasion.

Three excellent, highly detailed reports of the Little Rock trial have been published by major participants, expert witnesses on opposing sides. These reports are found in writings by Norman Geisler, Langdon Gilkey, and Michael Ruse. Geisler is a giant of the creationist movement and a brilliant Bible scholar who has lectured throughout the United States and in 25 foreign countries. He has also been dean of the Southern Evangelical Seminary in Texas. Langdon Gilkey was an equally distinguished scholar, prolific writer, and professor of theology at the University of Chicago Divinity School. Flamboyant in his long hair and earrings, Gilkey, a leading spokesperson for the liberal and ecumenical approach to Christianity, strongly opposed any incursions of religion into the public schools and nurtured a particular animosity for the creationist movement. Michael Ruse, a philosopher and historian of science, gave the most dispassionate report of the three, despite his opposition to creationism. His sometimes ironic view of the proceedings was especially refreshing.

Though it lacked the circus atmosphere of the earlier event in Dayton, the Arkansas trial was covered by reporters from several foreign countries. Conservatives, with considerable justification, felt that the press coverage was biased rather than informative and did not succeed in clarifying the issues. The Arkansas press and other papers presented the trial as a combat between science and fundamentalist religious beliefs, when other issues were certainly at stake. These accounts widely ignored the fact that the bill itself had reaffirmed the principle of separation of church and state and they left the impression that born-again theological and historical views had challenged this separation.

Just a month before the trial began, interestingly enough, an NBC poll had been taken (November 18, 1981) that revealed that three out of four Americans believed that both the scientific theory of evolution and the biblical theory of Creation should be taught in the public schools. Yet press coverage failed to note that taxpayer control of the public schools was even an issue. Witnesses for the defense were asked irrelevant questions about their personal religious beliefs, despite repeated objections by the defense attorneys, and the papers were quick to identify these witnesses as "biblical literalists." Since Geisler had briefly acknowledged in pretrial depositions his belief that UFOs were Satanic manifestations, much was made of these occultist views in attempts to undermine his scholarly and logical testimony. *Time* magazine and other publications, unfairly, presented him as a UFO cultist.

Geisler felt the media misrepresented the trial in several ways:

(1) They failed to stress the solid credentials of the scientists who were witnesses for the defense, even though both the Court and the ACLU recognized them all as "experts." (2) They neglected to report the anti-creation bias of the ACLU witnesses, though many of these were active in organizations with an anti-creationist agenda. (3) the media usually failed to report that many pro-evolution witnesses agreed that scientific evidence for creationism should be taught in the schools. (4) The media omitted mention of the religious and philosophical beliefs of the evolutionists; most were either liberal, agnostic, atheists, or Marxists.[7]

Press coverage of the trial revealed several problems. The generally liberal bias of the contemporary U.S. media has been well documented, but there has also been a loss of distinction in much of the press between news reporting and editorializing. The testimony of W. Scot Morrow, an agnostic evolutionist who spoke in favor of the Arkansas law, was poorly reported and distorted even when it was reported. Associate Professor of Chemistry at Wofford College in Spartanburg, South Carolina, Morrow testified in favor of "openness of inquiry" and "fair-play for minority opinion in regard to controversial issues." Morrow felt that his solid testimony on behalf of the defense had been dismissed as a "diatribe" by the press.[8]

Geisler contended that the minds of schoolchildren should be open to various possibilities and that both creationism and evolution should be introduced in the public schools, without religious applications. Only outside the classroom could both creationism and evolution be fruitfully debated from religious perspectives. Either could be used to reach certain religious affirmations, and neither should be excluded from the classroom because both could be so used. Scientific progress, he reminded his audience, must allow for dissenting theories, as the careers of Copernicus, Galileo, and Einstein demonstrate.

The ACLU lawyers challenging the act were highly skilled, backed by strong legal consultants. According to media accounts, there were as many as 22 attorneys involved in some way in prosecuting the case. They understood the effectiveness of drama in the courtroom and also knew how to overstress sensational but irrelevant matters, such as Geisler's UFO statement. Their frequent references to the religious background of the defense witnesses was also not directly relevant to the issues involved.

Judge William Jay Overton, in whose courtroom the case was heard, has been criticized and even accused of direct involvement in an ACLU plot. Geisler felt these accusations were unfair, although he found the judge's biases clear. Overton was a theologically liberal Methodist layman whose mother, a biology teacher, was present in the courtroom. Her distain for scientific creationism was evident by her facial expressions and comments. Overton's Methodist bishop was also one of the first witnesses for the prosecution. Geisler felt the bishop's testimony should have been disallowed or else the judge should have recused himself from the trial. Furthermore, Overton expressed personal opinions at various points in the proceedings. Although there was a week of testimony from a PhD in science who insisted that creationism was as scientific as evolution and not necessarily based on the Bible, the judge referred to the creationist movement as "the Biblical view of creation."

On January 5, 1982, Judge Overton struck down Act 590 as unconstitutional. In his ruling, Overton demonstrated, according to Geisler, several areas of ignorance. Overton's remarks about the creationist movement and its dating were in error. He appeared to believe that Creation narratives and the flood stories were unique to the Hebrew Bible, when similar accounts are in fact found in most ancient Near Eastern cultures. And his statements about the scientific community were also sometimes in error. Geisler felt that the judge did not fully understand either the issues of philosophy or religion that were at stake. For example, a belief in a Creator does not necessarily constitute a religious belief, as Geisler effectively pointed out in his testimony. The ancient Greek philosophers believed in a First Cause but did not worship it. It is also true that Darwin's *On the Origin of the Species* itself presents a legal problem for the science classroom, according to Judge Overton's understanding of what should be permitted there, because the last lines of the book do refer, in almost religious language, to a Creator who is responsible for the first forms of life. Geisler concluded that the state, unwittingly, may actually have been in the process of establishing the religion of Humanism in the public schools, as unconstitutional as any establishment of Christianity, or Judaism, or Islam.

Not surprisingly, Langdon Gilkey's take on the Arkansas trial was quite different from Geisler's. Gilkey was especially interesting when writing about his personal experiences, as he did in *Creationism on Trial: Evolution and God at Little Rock* (1985). As a witness for the ACLU from December 7 through December 9, 1981, he was extravagant in his praise of the ACLU lawyers who prosecuted the case and equally pleased with the other defense witnesses. He observed, convincingly, that the Little Rock courtroom provided one of the greatest scientific seminars ever. For his part in the trial, the lawyers expected him to provide a careful definition of religion. As a disciple of the celebrated German American theologian Paul Tillich, who identified religion as "ultimate concern," Gilkey felt well qualified for this task. He was also expected to explain why Act 590 had the effect of establishing religion and why creation science was in its very nature religious. His final task was to explain why and in what way creation science is not a proper science.[9]

Gilkey testified that Act 590 filled him with horror. He was further repelled that a segment of his own Christian community in the second half of the 20th century would marshal an attack on "the most fundamental and pervasive theorem of modern science." Problems with the Arkansas law, as Gilkey saw it, were many. It required every science class in the Arkansas public school system to balance an established scientific theory with one that was untenable. It required that whenever the evolutionary model was taught, creationism had to be given equal time and emphasis. In examining the law, Gilkey concluded that it would not only establish religion in the public schools, but a particular form of the Christian religion that he personally found offensive. It would give students the impression that only two views of human origins exist. Gilkey, a serious student of world religions, found the neglect of multiple religious views an astonishing omission.[10]

Finally, and perhaps most important of all, Gilkey believed that a dangerous political agenda upheld the creationist movement.[11] During the 1980s, national magazines and liberal religious publications constantly warned of the Moral Majority, a group led by the Virginia Baptist pastor Jerry Falwell. This group combined what it regarded as old-time religion with conservative economic and political views and aspired to sway elections toward candidates who shared their positions. Although their methods were no different from those of other political interest groups, liberals regarded them as a special threat. While the Moral Majority was able to exert minor influence, some of it detrimental, on U.S. foreign policy, its power was exaggerated. Within a few years, Falwell died, and his Moral Majority largely vanished from the public arena.

Knowing Gilkey's penchant for verbosity, his lawyers warned him to stay on track and not wander off with too many extraneous pronouncements. He was also told to stay within his own stated area of expertise. His approach to the Bible, as a "neo-orthodox" theologian, was made clear at the beginning.

Gilkey's description of Geisler, whom he immediately recognized as his counterpart for the defense, is interesting. While acknowledging this opponent's wide acquaintance with philosophical and theological issues, he found in Geisler a strange "marriage of literalism and modern science." And Geisler made some tactical mistakes. In the pretrial deposition, he had unfortunately acknowledged his beliefs in UFOs and a personal devil who operated through them, using these vehicles as his "major, in fact, final, attack on the earth." Geisler claimed he knew personally at least twelve people who had been possessed by the devil. And, to make things worse, he had given as his source of UFO information *The Reader's Digest,* a publication derided by intellectuals as the epitome of lower bourgeois sentimentality and bad taste. Although Gilkey had to leave the trial before Geisler's testimony, it was he who alerted the prosecuting attorneys to this vulnerability, and he was further delighted when all the newsmagazines and newspapers picked up on this part of the deposition.[12]

Gilkey expressed his admiration of the lineup of religion, philosophy, and science scholars who openly opposed the law. The religious leaders included Roman Catholic, Methodist, and Episcopalian officials, members of both the white and African American communities. The philosopher Michael Rose gave, in Gilkey's judgment, an especially eloquent testimony, clarifying distinctions between scientific endeavor and religious pursuits. The first group of witnesses presented the case that creation science was not a scientific but a religious model from a particular sectarian point of view. They were followed by scientific expert witnesses who explained important theories and research in geology, biophysics, biology, and paleontology. These prosecution witnesses contrasted sharply with the activities and claims of the defense witnesses.

Francisco Ayala was an especially noteworthy witness. He was not only a world-renowned geneticist but a former Dominican priest with theological and philosophical competence as well (see chapter 4). A courtly European intellectual, he spoke in a passionate Latin manner. Even the court reporter pronounced his Spanish accent "perfectly lovely." In his testimony, Ayala stressed that creation science did not present testable hypotheses. He pointed out that species, defined as breeding communities, do change, with the change sometimes rapid and observable, especially in the case of bacteria. With fruit flies, species changes can occur so quickly that they can be charted in the laboratory. Ayala also made the point that government control of science could be devastating, as Stalin's Soviet Union had proved.

The strongest scientific witness for the prosecution, according to Gilkey, was the most famous of all: Stephen Jay Gould, the Harvard paleontologist known to general audiences for his successful popular science writings. He presented evidence to establish what he regarded as the fact of evolution and demonstrated how this evidence supported his own (disputed) theory of punctuated equilibria.[13] Because Gould was skilled at presenting scientific information to a lay audience, he was especially effective. Gilkey felt that Gould's testimony established definitively that creation science was not a valid alternative to evolutionary biology but a novel hypothesis that, if triumphant, would mean the end of science as a tested and unified structure.

Unlike Geisler, Gilkey was impressed by Judge Overton's conduct of the trial and by his determinations. Overton accepted the ACLU argument and ruled that creation science is a religious doctrine rather than a valid scientific theory. Its model, he ruled, is taken from Genesis, its proponents are traditional fundamentalists and conservative evangelical groups, and it is founded on biblical literalism and belief in the Christian God. Thus, Judge Overton concluded, Act 590 sought the establishment of a particular religion in the public school system, inevitably involving a government "entanglement with religion," sought to advance that religion, and, consequently, contravened the First Amendment to the Constitution.

It might well be observed as a postscript to the trial that Gilkey in most ways represented American mainstream denominational Protestantism. But what he did not acknowledge was that this mainstream was no longer the central American faith commitment. For several years, these established Protestant denominations had been losing membership, while fundamentalist, evangelical, and Pentecostal groups were expanding rapidly. Roman Catholicism was now the single largest religious denomination in the country, though it, too, was losing members, especially among Hispanics, to the holiness groups and to the free churches.

Gilkey made clear that his own motivations were not purely scientific. He greatly feared a takeover of education and society by what he regarded as "a demonic religious mythology," a force of "ideological imperialism" that challenged a free, industrial civilization.[14] The Moral Majority, which he defined as a marriage of right-wing capitalistic society with Christian fundamentalism, was his declared enemy. Writing in the 1980s, Gilkey did not foresee the imminent decline of the Moral Majority or understand that its strength was largely a media myth. He was by no means alone in his fears. A literature had been spawned, even a fictional one, best represented by Margaret Atwood's outstanding book *The Handmaid's Tale* (1985). (Though the fictitious Gilead of Atwood's novel was intended to represent the United States after the triumph of reactionary religious forces, what the gifted Canadian novelist really succeeded in describing most accurately was Afghanistan under the Taliban.)

Gilkey also feared a totally secular society with "no ultimate vision, no unifying symbols of reality, truth, and value at all."[15] He envisioned as the ideal a society where religious communities would both support and correct the wider cultural life. But if the creationists achieved their goal in science classrooms, he warned, they would soon invade other areas of curricula. No pagan literature or art would be tolerated in humanities classes, and history classes would be dominated by warped interpretations of events and predictions of end times. For all these reasons, he expressed pleasure with what had taken place in the Little Rock courtroom.

Still a third opinionated, yet somewhat humorous, account of the trial was given by the science philosopher and historian Gilkey so much admired, Michael Ruse.[16] Ruse's descriptions of trial participants were vivid: Gilkey was "a rather trendy and over-articulate theologian" who was usually found flying from place to place investigating new sects for the Internal Revenue Service. The fundamentalist ministers who supported the defense had, according to Ruse, "that over-groomed look which is their trademark." He accused the creationists of falsifying their information and misquoting legitimate scientific literature, labeling their entire enterprise "corrupt," supported by "sleazy" scientists who testified for the defense.

Ruse admired the prosecution lawyers from the high-level New York firm who worked pro bono, but he could not resist adding that they no doubt were enriching their reputations, establishing their firm as something more than merely a high-priced operation. Nevertheless, they performed brilliantly and used their witnesses convincingly. He found it worth noting that, at the end of the trial, the lawyers joined the expert witnesses who still remained in Little Rock at the local pub singing "Amazing Grace" and other evangelical hymns.

Ruse, always the philosopher, had many other things to say about creation-ist and intelligent design proponents, contrasting them with earlier theistic evolutionists such as Asa Gray and Teilhard de Chardin. These earlier theistic scientists generally accepted Darwinian concepts but believed that God, who keeps himself largely hidden, works undetectably behind the process. They acknowledged that the divine role could be known only by faith and did not claim scientific verification for their beliefs. The ID champions who were try-ing to get their ideas into the public schools, on the other hand, believed that science could infer God's existence, which does not have to be taken merely on faith. Ruse concluded that the battle had become so heated not because it was between science and religion, but that it was between two religions, each holding its precepts with equal fervor. The outcome of the battle was not yet clear. But creationism, he felt, was not likely to lose its following anytime soon, because it was so closely linked with American visions and fears of the future. Science itself was a religion with its own program for saving society and the world. Sympathetic cooperation rather than antagonism would be the best course for the future, though Ruse believed this to be unlikely.

The Pennsylvania Case

Another publicized, precedent-setting courtroom drama unfolded in Dover, Pennsylvania, in 2005. This time, intelligent design rather than scien-tific creationism was in the dock, and it was widely believed that opponents of creationism would be presented with a more sophisticated defense than they had previously encountered. ID champion William Dembski voiced the opinion that the Dover trial would provide a grand opportunity to "squeeze the truth" out of the Darwinists. Even more than the Arkansas trial, Dover provided a forum for distinguished expert witnesses from around the country to present what amounted to a first-class seminar in science and religion.

The plaintiff in the Dover case was Timmy Kitzmiller, and the defen-dant was the Dover Area School District. It had all started in October 2004, when the school board adopted a resolution to the effect that students were to be made aware of gaps and problems in Darwin's theory and that other theories of evolution, including but not limited to intelligent design, were to be acknowledged.

In order to respond to this mandate, the school board decreed that all ninth-grade biology classes at Dover High School would be read the following statement:

The Pennsylvania Academic Standards require students to learn about Darwin's Theory of Evolution and eventually to take a standardized test of which evolution is a part. Because Darwin's theory is a theory, it continues to be tested as new evidence is discovered. The Theory is not a fact. Gaps in the Theory exist for which there is no evidence. A theory is defined as a well-tested explanation that unifies a broad range of observations. Intelligent Design is an explanation of the origin of life that differs from Darwin's view. The reference book, *Of Pandas and People*, is available for students who might be interested in gaining an understanding of what Intelligent Design actually involves. With respect to any theory, students are encouraged to keep an open mind. The school leaves the discussion of the Origins of Life to individual students and their families. As a Standards-driven district, class instruction focuses upon preparing students to achieve proficiency on Standards-based assessments.[17]

What would have been regarded in calmer times as an innocuous statement designed to placate a number of parents, while leaving explorations of origins up to interested students and their parents, caused immediate dissension. This was too strong a concession to religious sentiments to please a number of citizens and the organizations that supported them.

In December 2005, the District Court of the Middle District of Pennsylvania ruled that the Dover School District policy was unconstitutional, that ID and creationism (its progenitor) were not scientific and should not be taught in classrooms, that they were basically religious views and, consequently, in violation of the Establishment Clause.

A vivid account of the trial has been given by participant Kenneth R. Miller.[18] Miller—an evolutionary biologist, author of popular science books, and a practicing Roman Catholic—was a lively witness. His testimony dominated the first two days of the trial, presenting the scientific case for evolution.

Miller still believes that the Dover trial was a decisive battle for the ID movement, which failed to convince the court of the validity of any of its arguments. Eight proponents of ID, including Dembski, had agreed to appear as expert witnesses in defense of the Dover school board. Only three, not including Dembski, actually testified. Miller contends that the others lost heart, though they claimed conflicting interests and insufficient legal council. The result was that the ID movement was not as strongly represented as it might have been, and the prosecution witnesses were easily able to respond to ID arguments. The honesty of ID proponents also came under attack, as they were accused of consistently obscuring the religious motivations of their movement and lying under oath by denying these aims.

All the standard arguments of ID had, to the satisfaction of the court, been refuted by the plaintiff's expert witnesses. Michael Behe's favorite examples of irreducible complexity on close inspection were said to be deficient. Natural selection could and often did recombine features of one system with features of others for different functions. Genes could be duplicated, changed, and amplified by natural selection. Expert witnesses all agreed that gaps in evolutionary knowledge did not authenticate ID, because science was constantly closing these gaps. ID could not be verified or falsified by scientific methods but was grounded only in the authority of religion, the court concluded. Therefore, it had no place in the public school curriculum.

Kenneth Miller was heartened by the outcome of the trial. ID defenders had failed in their attack on evolution in two principal ways. They had been unable to deal with the enormous scientific evidence that supports evolution, and they had failed to provide anything that qualified as an alternative theory to it.

NOTES

1. In *Reynolds v. United States* (1879), the Supreme Court found no constitutionally compelling exemption for Mormon polygamists. While the government cannot regulate religious beliefs, it concluded, it can regulate religiously motivated conduct that violates community standards. The case grew out of attempts by the administration of President Ulysses S. Grant to eliminate polygamous practice in lands controlled by the U.S. government.

2. In *Sherbert v. Verner* (1963), the Supreme Court ordered a state to pay unemployment benefits to a Seventh-Day Adventist even though her refusal to work on Saturdays otherwise would have disqualified her from such benefits. In *Wisconsin v. Yoder* (1972), the Court determined that Amish were not required to send their children to school after eighth grade. In *United States v. Lee* (1982), the Court refused to grant Amish exemption from social security taxes.

3. *Lemon v. Kurtzman,* 403 U.S. 602 (1970).

4. See the books by Edward J. Larson recommended in the Bibliography.

5. This song was composed by Uncle Dave Macon in 1926. It was transcribed from Vocalion 15322 and 5098. See Charles Wolfe, "The Good Book in Country Music." For a fuller look at country music's response to the Scopes trial, see *The Bible and Popular Culture in America,* ed. Allene Stuart Phy (Philadelphia and Chico, CA: Fortress Press and Scholars Press, 1985), 93–95.

6. John T. Scopes and James Presley, *Center of the Story: Memoirs of John T. Scopes* (New York: Rinehart and Winston, 1967), 276.

7. Norman Geisler, *Creation and the Courts* (Wheaton, IL: Crossway Books, 2007).

8. Ibid., 223.

9. Langdon Gilkey, *Creationism on Trial: Evolution and God at Little Rock* (San Francisco: Harper & Row, 1985), 7.

10. Ibid., 9.

11. Ibid., 41.

12. Punctuated equilibria is a theory developed by Stephen Jay Gould and Niles Eldredge to explain the Cambrian explosion and similar phenomena. According to this theory, evolution proceeds in fits and starts, punctuated by long periods of stability.

13. Gilkey, 76.

14. Ibid., 203.

15. Ibid., 221.

16. Michael Ruse, "A Philosopher's Day in Court," in *Science and Creationism,* ed. Ashley Montagu (New York: Oxford University Press, 1984), 311–342.

17. Quoted in Geisler, 214.

18. Kenneth R. Miller, *Only a Theory: Evolution and the Battle for America's Soul* (New York: Viking, 2008), 70–71, 73–74, 178–179, 208–212.

8

Public Schools and Religion

Though scientific creationism and intelligent design have been banished from classrooms, the courts have reaffirmed in their rulings the need for instruction about religion as part of a program of secular education in the public schools. Yet because so many restrictions surround religion, schools have found it much safer simply to ignore the subject. The result is an impoverished educational program that leaves important questions hanging. How can these schools, despite all the problems and prohibitions, best inform their students of the role that religion has played in world history? How can teachers and curriculum planners function when they must constantly walk a tightrope between fundamentalist and atheist parents, between evangelical churches and the American Civil Liberties Union? The task is daunting but cannot be neglected.[1]

Because of the perceived hostility of the courts to religion, many observers of contemporary U.S. public schools feel that the system's very survival may be at stake. Too many students are leaving public schools, which treat religion as if it has never been a central part of human life. Regardless of personal faith or the lack of it, this picture of a totally secular world is inadequate. Of course, these same critics may resent the impartiality that a public school teacher must show toward religion, or they may not understand why religion may not be favored over nonreligion. In addition to recognizing people of faith and their many contributions to human civilization, the courts have made clear that free thinkers, skeptics, and spiritual iconoclasts also have to be given voice.

Relatively few teachers are qualified to direct student discussions and activities in the subject of religion. Some do not clearly understand how a matter so emotionally fraught can be presented objectively. Although many teachers may have an adequate understanding of the religions in their communities, not many of them will have had the opportunity to learn about Middle Eastern and Asian religions, though these play increasing roles in international affairs and have a substantial presence in many American and European communities.

Because religion must be handled so gingerly in the classroom and the courts have done so much to make Bible reading and any actions that may appear to promote religion difficult, teachers may not always clearly understand what may and may not be said on the subject. Considering all the delicate sensibilities that may be offended, as well as the ways a teacher's words may be misconstrued and misreported to parents, it often seems easier just to ignore the subject. But the result will certainly be a one-dimensional view of history, literature, the arts, and even current affairs.

Many educators may not initially understand the importance of studies about religion. They will probably have been educated in secular colleges and universities where religion has been largely ignored. If their studies have concentrated on the sciences, which do not resort to the God hypothesis, they probably have spent their time making sense of the world without recourse to theological answers. Furthermore, scientific associations and educational organizations, in responding to scientific creationism and intelligent design, have made known their desire to have no part of religion in science classrooms.

Teachers who have majored in the humanities will probably be better equipped to deal with religious topics. Some teachers will have received their training in colleges sponsored by religious denominations and will probably be better informed on the subject, but they may have had less opportunity to interact with those who do not share their specific beliefs. They may, consequently, find objective discussions of religious subjects difficult.

Whatever the reasons, when a conscientious teacher decides to include instruction about religion in the educational program, the first task will be to determine exactly what religion is. Some religions stress uniformity of belief, orthodoxy, framing creeds and developing elaborate theologies. Historic Christianity, with Islam to a lesser degree, are examples. Other religions care less about orthodoxy of belief but demand a particular style of life, often a set of ceremonial or ritualistic rules as well as standards of ethical conduct. Judaism is a good example of this approach. Still other religions are less concerned with the intellectual or social content of faith than they are with achieving a spiritual bond between themselves and the divine. Hinduism, with slight oversimplification, offers such an example. Religions like Buddhism avoid speculating on the nature of God or the gods but assist their faithful in finding ways to endure or transcend the perils of the human situation. Classical Confucianism, the traditional faith of ancient China, is essentially a social philosophy. While a Paul Tillich may define religion as "ultimate concern" and speak of God as "the ground of all being," most people look for more specific definitions and directions. Certainly students will ask for more concrete examples.

Teachers in tax-supported schools, while impartial in their presentations and avoiding any attempts at proselytizing, must still decide how much classroom time should be devoted to instruction about particular religions. Will they concentrate on faiths that touch most directly the students under their charge? Or will they identify religions that have exerted the greatest influence on world civilization, whose lore must be known if one is to fully appreciate literature and the arts? Will they tell the story of minority religions, such as Mormonism, which has been so important to the United States and has a history that is lively and dramatic? Will the teacher be well enough informed on the rich and unique traditions of African American spirituality to do them justice?

Teachers must note as well the cultural wars that are now going on in American society. Religion plays a part in them, and students need to have some understanding of the difference between liberal, conservative, evangelical, and fundamentalist factions, difficult though these are to define. They should know that these are movements that cross denominational lines. In order to understand the major ethical-religious debates—over stem cell research, abortion, euthanasia, cloning, gay marriage, and so on—they need to have some understanding of why radically different positions on these issues may be held by equally sincere people. U.S. history teachers need to convey knowledge of the role religion has played in such movements as abolition, prohibition, women's suffrage, and civil rights. In the past there have been relatively few resources available to teachers wishing to give proper attention to religious topics in public schools. That is now changing as more educators realize the importance of the subject and as the courts clarify what is appropriate in the public schools.

RELIGION'S PLACE IN THE CURRICULUM

If an individual teacher or a school decides to include discussions of religion in the classroom, attention moves quickly to where and at what grade level this should take place. If it is decided that religious studies should be integrated throughout the curriculum, interdisciplinary team teaching has much to offer and is growing in popularity. If special classes in religious topics are to be given, teachers need the proper training in both the subject matter and the law.

When school systems conclude they have the time and resources, special classes in world religions are usually offered in high schools, often in honors programs or in college preparatory and advanced placement classes. At the minimum, the teacher should be prepared to discuss ancient paganism, Hinduism, Buddhism, Confucianism, Taoism, Judaism, Christianity, and Islam. Some programs have eliminated Judaism and Christianity on the mistaken assumption that American students will already know the basics of these faiths. A teacher should also be able to relate not only historical movements but current events to the faith and practice of each religion. Ethnic conflicts in India, for example, may be discussed, including major events such as the partition of Pakistan from India. The concept of jihad, or holy war, as it has historically been implicit in the Hebrew Bible, during the Christian crusades, and in early Islam, is another important topic. The Armenian genocide, the Nazi Holocaust during World War II, the more recent massacres in parts of Africa, and the conflicts between Catholics and Protestants in Ireland would be other topics to examine.

But while a class in world religions would be valuable to college-bound students, even the average student should leave high school with some awareness of the role of religion in history and current events. This is where the argument for the integration of religious issues into all high school classes may be made. The study of history, political science, sociology, psychology, literature, and art is incomplete without the acknowledgement of religious themes. It is worthwhile for students to understand how religion has permeated almost every activity, how it has influenced the legal systems, and how it has been the basis of most holidays and many customs. While the teacher or discussion leader must always strive for objectivity, it is also appropriate to point out the failures as well as the achievements of religions. To present only the favorable aspects of world religions—with no mention of the conflicts they have generated, the inequities they have sometimes countenanced, or the bigotry that has sometimes been expressed—would be as wrong as to present only negative features. This is the point at which impartiality becomes especially difficult to maintain. There is always the temptation to compare the grand ideal of one's own religion with the common practice of the faithful in other religions.

Ideally, world religions should be presented descriptively by a well-informed teacher who explains beliefs and practices. But this never provides a full picture. It is also desirable to have guest speakers who practice these religions, especially those that seem exotic to students. Field trips are helpful, despite the lengthy preparations and time required to obtain all the proper permissions and set up the schedules. They allow students to interact with leaders and communicants of different faiths. These activities, useful as they are, can be precarious, because it still must be made clear that no proselytizing is taking place.

The use of both primary and secondary resources should be encouraged. Different translations of the Bible, copies of the Qur'an—which, according to Muslims, cannot really be translated but only approximated in languages other than Arabic—the Hindu classics, the Book of Mormon, and so on should be available in school libraries. Even strong students will require some instruction if they are to understand ancient documents from unfamiliar cultures. Most students will read only snippets, but even this will give them some of the flavor of the holy books. Oral readings help bring these documents to life. Even a recording of the Qur'an chanted in Arabic by schoolchildren will reveal some of the power of this holy book. The very concept of sacred scripture should be discussed, noting again how important the development of writing has been to the human race, providing the ability to preserve the experience and the emotions of those who have gone before. Is it, therefore, surprising that many people have believed that the gods themselves have chosen to deliver promises, commands, and admonitions through holy writings?

Although more analytical classes in religion are appropriate to secondary education, even younger children in the elementary grades will enjoy some of the lore of the great world religions. Bible stories, episodes from the life of Muhammad, the Jataka tales of Buddhism, stories from the two great Hindu epics (the Ramayana and Mahabarata), and tales of the Arabian Nights are enjoyable and introduce children to the diversity of world traditions. The Homeric tales, those classics of paganism, in beautifully illustrated adaptations appropriate to children, have long been used in elementary classrooms.

Of course, a teacher must be prepared to answer objections from parents who do not always understand the uses and values of creative literature and are anxious about the possible indoctrination of young children. Some parents, like a few children themselves, cannot always separate flights of the imagination from factual reportage and find tales of monsters, demons, and legendary heroes a waste of valuable school time. It should be remembered that, while the more capable high school students can deal with ambiguities and different points of view, very young children usually cannot. Their parents sometimes cannot, either. The preparation of teachers is of special importance here. All too often, teachers must spend too much time in college taking courses in educational theory, completing projects that have little to do with subject matter. Insufficient time is left to study the subjects they will be called upon to teach.

Some students get little information from their parents, and many are unchurched. But they still need to know how important religion has been to history and civilization. And even those children who are regularly sent to Sunday schools in their parents' churches rarely receive any systematic instruction that can be easily integrated into the other subjects they are learning. Young people who attend church regularly and have been enrolled in Sunday school since childhood may still have little understanding of the traditional theology of the Christian churches or even the tenets of their own denomination. Their knowledge of church history may be nonexistent, and their study of the Bible, if it has existed at all, has often been so unsystematic and haphazard as to be almost useless. Even in the American Bible Belts, today the state of biblical literacy is grim.

It can be suggested that such knowledge is unnecessary in a secular age, and this is no doubt the way many students, even those from religious homes, on some level feel. However, there is some loss when our common frame of cultural reference becomes what we have heard on Oprah Winfrey's or another popular television program. Matthew Arnold's concept of the educated person, who cherishes what the finest minds have preserved from the past and projects it forward to the next generation, often seems old-fashioned.

Some of the most dramatic episodes in U.S. history have been motivated by religion. The struggle between Old World Christian establishments and Quakers, Shakers, and Puritans is a vivid part of American history. The life and martyrdom of Joseph Smith and the entire history of Mormonism is a distinctively American epic. Many times, Americans have given voice to their sense of destiny, which, for better or for worse, has been grounded in religious belief. As Warren A. Nord and Charles C. Haynes write in *Taking Religion Seriously across the Curriculum*:

Some multiculturalisms disparage . . . principles and values of the American republic and western civilization as "Eurocentric" and oppressive. Although we agree that teachers should discuss both virtues and vices of the West and the United States, we would argue that public schools have an obligation to teach and uphold the democratic first principles of the U.S. Constitution with its Bill of Rights. Yes, many of these principles are derived from European sources and from the biblical traditions. (And these roots should be taught.)[2]

But religion is not merely a relic of history. Even in the contemporary world, an awareness of religious movements and developments is essential. The last two centuries cannot fully be understood by ignoring religion, even when it seems that trends have—in most ways—been secular. The debates within Roman Catholicism during and since the Second Vatican Council have changed the way religious groups deal with one another in American society. The enormous growth of Mormonism and the influx of Muslims from the Middle East have also changed the U.S. landscape.

We usually think of literature and the arts as the areas where religion has exerted the strongest influence, but we should not forget the enormous theological literature that deals with economic issues. Why did the communist establishment of the Soviet Union fear and distrust religion so totally? Why did communist China try to replace the Analects of Confucius with the Little Red Book of Chairman Mao? What do the churches today have to say about world poverty, hunger, and the worldwide use and abuse of natural resources? What about the environment and economic growth? Which religions have made work a religious duty, and which have dismissed it and other human activity as immediately unavoidable but ultimately meaningless?

And it should never be forgotten that the impetus for education and culture throughout the world has chiefly been derived from religion. Buddhism and medieval Christianity produced the two greatest bodies of visual art in the world. In early Judaism and Christianity, dance was associated with pagan religious rites and became the least developed of the arts in Europe, while it flourished in Asian cultures.

The intrusion of religious issues into academic subjects is most forcefully opposed in the sciences. But historians of science, as we have seen, freely acknowledge that the two disciplines of theology and experimental science have frequently worked together in the past. Religion has been entangled in every human endeavor, in ways that cannot be ignored in education. Scientific creationists and intelligent design champions want religion back in the science classroom, while their opponents just as strongly believe that one of the aims of education should be to free the minds of students from the prejudices and superstitions of the past, to liberate them from oppressive forces of religion. This is an intellectual battle that will not soon end.

NOTES

1. The following books are especially recommended: Charles C. Haynes and Oliver Thomas, *Finding Common Ground: A First Amendment Guide to Religion and Public School* (Nashville, TN: First Amendment Center, 2007); Warren A. Nord, *Religion and American Education* (Chapel Hill: University of North Carolina Press, 1995).

2. Warren A. Nord and Charles C. Haynes, *Taking Religion Seriously across the Curriculum* (Nashville, TN: First Amendment Center, 1998), 87.

9

Science and Religion, Now and in the Future

A half-century ago, British novelist and scientist C. P. Snow gave his germinal, yet even-then controversial lecture, later published in *The New Statesman,* on the relationship between science and the humanities. At that time, he coined the phrase "the two cultures." Snow felt he had some direct knowledge of his subject, being a man of science by training and an imaginative writer by trade. His observations in many ways seem as pertinent today as they were when he first made them and are of special relevance to education.

THE TWO CULTURES

Snow argued that education is inadequate, if not actually dangerous, when it becomes too specialized. Since Snow wrote, we have moved more and more into an age of specialization, and it seems scientists and humanists not only speak separate languages but hardly communicate with one another at all. This barrier is especially high between religious and scientific people, and educators are caught in the middle of the struggle. Both groups accuse the other of contributing to the ultimate horrors of the 20th century: weapons of mass destruction, Auschwitz, and environmental disaster. Scientists point to the fascist leanings of such literary figures as Ezra Pound, William Butler Yeats, Wyndham Lewis, and others, while humanists blame the eugenics applications of remorseless science for the racism of the Nazis.[1]

Snow, who straddled the two cultures in his own work, was somewhat more optimistic than intellectuals today. He thought science would solve the disparities between developed and Third World countries, would find ways to feed the earth's entire population, and would halt the spread of communism. His predictions for the third millennium, like most such predictions, turned out to be inaccurate. Nevertheless, his observations are still worth consideration.

In Snow's view, the traditional culture, that which he saw perpetuated by the great universities of Oxford and Cambridge, is basically literary. Yet he believed traditional culture is constantly losing ground, growing defensive, and increasingly resting on its "precarious dignity." The scientific culture, in the meantime, expands, growing confident as it gains in authority, solving one mystery after another while providing solutions to real problems. Scientists earn public support, while the other culture relies on an aristocratic elitism. Neither culture feels the other is worth knowing.

Using a literary man's metaphors that would be objectionable today, Snow described the scientific culture as "steadily heterosexual," lacking anything that may be described as "feline or oblique." Scientists, he allowed, are ready to accept history to a degree, especially relishing social history. Biography is their preferred reading, when they read at all. Philosophy, however, is rejected as irrelevant. Scientists, who conduct experiments and achieve results, it might be added, would probably agree with those who complain that philosophers have argued the same questions for centuries without reaching any conclusions!

Snow further observed that scientists tend to be indifferent to the arts, with some exception for music. Architecture is concrete enough for them to enjoy, but the more abstract arts, again with the exception of music, are difficult for them to appreciate, while poetry is regarded as effete and useless. In literature they may occasionally relax with a light adventure novel—science fiction being a favorite—but arty writers such as William Faulkner or Marcel Proust annoy and bewilder them. While they respond to a Neville Shute (who wrote the apocalyptic novel *On the Beach,* about the world's destruction by atomic weapons), their notion of an experimental novelist is Charles Dickens, approaching him as if he were James Joyce (the Irish writer who said he expected his readers to devote a lifetime to his difficult prose). Scientists are pragmatic. They believe books should convey information rather than provide some nebulous aesthetic pleasure.

But Snow regarded this lack of communication between scientists and creative artists as lamentable. A life, he believed, is impoverished if a person has not read Tolstoy, Stendhal, or Balzac, while an individual certainly lives an unexamined life if ignorant of the basic laws of thermodynamics. Even psychoanalysis and cybernetics, Snow suggested, might turn out to be important!

To the surprise of many who believe science is amoral while ethics is the province of religion and philosophy, Snow concluded that scientists have made the major contribution to the moral enrichment of modern life. The novelists—and he might have provided many examples from the major literatures of the world—and sometimes even the philosophers have come to accept with a sort of bitter joy the tragic nature of the human condition. After all, has not tragedy been designated the highest form of literature? Do we not, he might have added, most revere from ancient writing the great tragedians—Aeschylus, Sophocles, Euripides—and have not modern-age writers such as Nathaniel Hawthorne and Henry James delighted in depicting picturesque poverty and gloomy wrongs? Poets such as the Frenchman Paul Claudel talk about the virtue of suffering, especially if it is the suffering of others, and during the 1950s, the Mississippi novelist William Faulkner provided sentimental reasons for the second-class citizenship of African Americans. Artists prefer a structured society in which people stay in their place, especially if their own place is privileged. Scientists are blessedly free from this conceit; their culture detests "defeat, self-indulgence, and moral vanity."

Snow acknowledged that artists may have found their ultimate inspiration in the tragic vision of life, but scientists have been busy finding solutions to poverty, drudgery, sickness, and early death. Instead of celebrating the endurance of tragic heroes, scientists have attempted to limit the human tragedy as much as possible, improve life, provide people with enough leisure to savor art, and make existence more comfortable and productive.

Snow, who made his points so eloquently, might have delved further into the mindset of humanists and scientists. He might have observed that the general public has unfavorable stereotypes of both. While the absent-minded humanist in his ivory tower is believed to resist intrusions from the real world, clothing his thoughts in golden words that obscure reality, the mad scientist, in the popular mind, is at work creating Frankenstein monsters, atomic bombs, and designer babies, searching for knowledge at any cost, without regard to the possible horrors of its applications. While Ezra Pound was broadcasting his anti-Semitic tirades in Italy during World War II, Josef Mengele was performing horrifying medical experiments on concentration camp inmates.

Snow's observations are also pertinent to the Darwin–intelligent design controversy. According to some surveys, a plurality of scientists fail to acknowledge the value of the intellectual traditions of Judaism, Christianity, or other religions. They are oblivious of the Islamic Golden Age. They deny the very religious foundations of the sciences they pursue. Even as serious theologians are trying to reconcile revealed truth to scientific observation, the masses of people find Darwinism repellent. Still, they readily accept the medical benefits developed from this science. The very anger with which Darwinists and intelligent design champions conduct their controversies over the heads of schoolchildren is witness again to the fact that they no longer truly communicate with one another. For modern folk, the two traditional books outlining the work of God—the Book of Scripture and the Book of Nature—contradict one another at every point.

PHILOSOPHICAL NATURALISM

In the postmodern period, both religion and science are under attack in many intellectual circles, at the same time intensely partisan religious movements are revived in many parts of the world. Regardless of language, country, and religious heritage, numerous nations are currently experiencing deep cultural clashes that too often result in violence. While masses of people are intent on defending religious views, an intellectual elite has embraced a philosophical naturalism. Are schoolchildren to be plunged into the midst of this battle?

One of the clearest, most perceptive statements of the philosophical naturalism that touches so many educated people today came from a British civil servant and author of highly accessible books on philosophy. In 1961, W. T. Stace published *Religion and the Modern Mind*, with *Mysticism and Philosophy* appearing the following year. Stace acknowledged that in earlier times people—and he was speaking of the intellectual minority—believed that the more we learn about nature, the better we can understand God's laws. The laws of nature were understood as God's way of getting things done, there for humans to discover in order to better their lot. The essential goodness of nature was widely affirmed.

However, a change began in the 17th century to be culminated in the first half of the 20th century, brought on by two destructive world wars, violent class struggles, and economic woes. Even the essential goodness of nature was now denied in a clearer recognition of the painful struggle of all living things to survive. Both events and scientific discoveries played their part in a growing pessimism. Stace clearly saw that natural philosophy had turned pessimistic, in sharp contrast to the optimism that had characterized much of it during the late 19th and early 20th centuries. Certainly the human lot, at least in industrial countries during peacetimes, had benefited from the increasing knowledge of the material world and the universe. But this knowledge had not made people happier.[2]

The mood about which Stace wrote has, he believed, much to do with the rejection of traditional religions throughout the world, a fact that he may well have exaggerated, as later events suggest. But he was not speaking of the opinions of the masses but of an educated elite, the movers and shakers of culture whom he saw no longer accepting the faith of their fathers. Ideals now had no transcendent foundation but were understood to proceed only from human minds. Standards of conduct were merely the fallible responses of humans to social situations. A spiritual emptiness could be felt everywhere, as men and women looked out on a dead universe, indifferent to human needs, values, and aspirations.

It is definitely science, Stace asserted, more than anything else that has brought about this radical change in perspective. This is not the result of a particular scientific theory. Religion could survive, with some modifications, Darwin's theory of evolution or the discoveries of geologists about the actual age the earth. Only details would have to change to bring faith in line with these discoveries. Science itself could not shake the basic dogmas of Christian faith: the Incarnation, the Resurrection, the Atonement. It is, rather, the general spirit of science and the basic assumptions on which the scientific endeavor rests that have proven incompatible with any religion. This started, according to Stace, in the 17th century, when scientists rejected any concern with "final causes," the belief that the universe is guided by some overriding Intelligence. All of Western civilization—whether pagan, Jewish, Christian, Islamic, or Deist—had earlier agreed that there is design and direction in the universe. This shift began unwittingly with the founders of modern science, with Galileo, Kepler, and Newton, devout as they were. They introduced the habit of ignoring any inquiry into final causes and examined instead ways to understand the material world and to predict and control events.

The result, certainly unforeseen and unintended by Galileo, Kepler, Copernicus, and Newton, is a view of the universe as "purposeless, senseless, meaningless." Nature is only matter in motion, with matter governed by blind force. According to Stace: "Religion can get on with any sort of astronomy, geology, biology, physics. But it cannot get on with a purposeless and meaningless universe."[3]

With this very modern perception that ultimately everything is irrational, the ancient philosophical-theological problem of the existence of evil vanishes. Pain and suffering have no significance, no value; they just happen in a senseless universe. Stace found this meaningless world reflected in the chaos of modern art, in discordant music, in abstract, surreal paintings, even in contemporary fiction. Today it would be tempting to add to Stace's own examples the theater of the absurd or a work of literature like William Faulkner's *The Sound and the Fury*, in which a part of the narrative is literally "told by an idiot."

An especially sad result of the collapse of religious vision, Stace went on to observe, is the disappearance of moral principles except as a set of practical though tentative rules. If God is dead, then, as a character in Fyodor Dostoevsky's novel *The Brothers Karamazov* announced, all is permissible. There are no God-delivered rules, and our morals merely express our personal likes and dislikes or the current prejudices of society. Philosophers, of course, find ways of defending standards of conduct that make civilization possible. Stace acknowledged that men like Thomas Henry Huxley, Bertrand Russell, and John Dewey might be able to lead productive and exemplary lives on the basis of a secular morality of their own concoction. But the masses of people find it difficult to forgo immediate pleasures without a belief in a God to love and fear, for whom they are ready to delay gratifications and make sacrifices.

Stace noted that several solutions to our present perplexity have been suggested by philosophers and social theorists. A genuine secular basis for right conduct might be worked out and sold to the average person. However, cool intellect alone could hardly sustain moral conduct. Yet as we face the bleak present and the uncertain future, there are other possibilities. Perhaps there will be a return to strong belief in God and the doctrines of a major religion such as Christianity or Islam. In some places, since Stace's time, this seems already to have occurred. Stace himself found such a return to orthodoxy unlikely, because he believed the religious vision has been destroyed. During the French Revolution, some thinkers indeed concluded that if God did not exist, he would have to be invented for the common folk. But, especially in an age of mass media, could an invented religion be sustained? Would it not mean an unhealthy manipulation of ordinary people by charlatans or an educated and sometimes ruthless elite? And if, in the improbable event that it succeeded, would it not be reprehensible? What power it would give to those who made gods and lawgivers of themselves! Humans have never been very successful in calmly constructing religions on strictly intellectual bases, without revelations, miracles, inspiring art, and the perception of crisis situations that demand the promise of some deus ex machina.

There is still a third possibility that Stace suggested: the emergence of a sincere new religion, grounded in what we know of science yet with its own inspiring vision. Perhaps the new faith would be led by an exciting, messianic personality with a convincing message, somewhat different from that which has been lost but still affirming of human life. Though Stace did not suggest it, we might today look at the worldwide success of faiths like Bahai and Mormonism, which originated with charismatic personalities in the modern world and in the full light of history. Whether these newer religions are compatible with science and modern views of gender equality is open to discussion.

Religions, Stace might have added, no matter how admirably constructed, appear never to have worked if their only intent has been to be practical, psychologically affirming, and socially sound. A religion to succeed has to emerge from some deep-seated needs, rarely understood by those who accept the faith, and be spread by leaders who, consciously or subconsciously, respond to these needs, which even science cannot as yet meet or explain. Stace concluded that "those who talk of a new religion are merely hoping for a new opiate."[4] Human beings have lived by illusions, by fables and dreams, by myths of love and grandeur. Perhaps they can keep some of their lesser illusions, but thinking people have now, he felt, been deprived of their Great Illusion:

Can he [modern mankind] grasp the real world as it actually is, stark and bleak, without its romantic or religious halo, and still retain his ideals, striving for great ends and noble achievements? If he can, all may yet be well. If he cannot he will probably sink back into the savagery and brutality from which he came, taking a humble place once more among the lower animals.[5]

SCIENCE IN THE POSTMODERN ERA

Though science gained in prestige throughout most of the 20th century, near the end of the second millennium and into the first years of the 21st century, like religion, it has faced increasing skepticism and new challenges. This has come from the postmodernists and deconstructionists, who cast doubt on all old verities, traditions, and canons. While scientists luxuriated in their realism—their belief that there is a real world quite apart from our fancies, beliefs, and hallowed traditions, a world that can be at least partially known by objective methods—these new thinkers have been asserting that any exterior world is always beyond human grasp, since human beings are imprisoned by the languages and symbols they use. Reality is always filtered through these signs and symbols.

Just as scientists attempt to propose theories that accurately account for a real world, the deconstructionists protest that science perhaps as much as literature and religion—and all the humanities—is just as dependent on values and interests. They like to point to the scientific theories of the past, happily swept away, that promoted racist, sexist, and imperialist doctrines. When we feel science has moved beyond this, postmodernists suggest that we are further deluding ourselves. We still have questionable scientific studies that suggest different levels of intelligence among races or genes that predetermine sexual orientation or spirituality.

Scientists maintain that while their theories always have a tentative feature, ready to be corrected or improved by new developments, they do correspond to the world and the universe that can be perceived. Science, although subject to the limitations of the human mind and senses, does concern itself with objective reality, unlike imaginative literature or religion. Postmodernists counter again that in the history of science, theories have changed so drastically that it is the ultimate delusion to assert that they are in any measurable way moving toward a greater grasp of reality, whatever that may be.

Science again, in contrast to the humanistic disciplines, operates through its tested method, and its practitioners correct themselves when their previous conclusions prove inadequate. The chief methods are reproducibility, consistency, falsification: the scientific method. Furthermore, findings are peer reviewed; other scientists are only too ready to point out errors and limitations, offering correctives. Again, postmodernists contend that scientists work within an orthodoxy. Their doctrines guide all their attempts at proof and verification, forcing them to operate upon a faith in their established procedures and premises.

While scientists readily admit that they operate through assumptions and paradigms honored within their disciplines, and many are ready to acknowledge that they too conform to a system of values, they believe that they—unlike religionists who hold to rigid orthodoxies—are always open to new discoveries and corrections that expand their awareness of the material world. They also remind their critics again that one of the best ways to establish a major scientific reputation, which all of them seek, is to provide decisive proof than an older received scientific theory must be replaced by a newer one. Postmodern critics of science are not convinced. They insist that scientists like humanists remain restricted by inherited assumptions, and they are equally enslaved mentally to their cultural time and place. Scientific language, like the language of the humanists, even like the language of everyday discourse, is metaphoric, viewing reality through a darkened glass.

RELIGION IN THE 21ST CENTURY

Despite attacks from several directions and numerous prophecies of its demise made a century ago, religion survives into the 21st century. With the ascendancy of science, many intellectuals believed that religion would disappear from the developed parts of the world. Soviet states placed restrictions on religious instruction and, equating religion with superstition, predicted such faith would slowly but certainly disappear from communist lands. Now, on the contrary, religion has outlasted communism in most countries and even shows signs of renewal in parts of the world. It remains a dominant influence in human lives today just as it has been in the past. When we study ancient civilizations such as the Egyptian, Babylonian, Sumerian, Hebrew, Greek, and Roman, we see how religious aspirations, beliefs, fears, and ceremonies have been at the heart of the development of all the arts and have strongly influenced laws and customs. Dare we deny that Western science itself is built on an intellectual foundation created by the Judeo-Christian-Islamic view of the world?

A recognized province of religion, at least in the West, has long been morals and ethics. Judaism, Christianity, and Islam are ethical monotheistic religions. Yet during the last half of the 20th century, a social revolution in what is deemed morally and ethically acceptable has taken place throughout the Western world. Standards of conduct previously viewed as scandalous, even in secular circles not to mention religious ones, became acceptable in almost all levels of society. Changes in family structure are perhaps the most evident examples of this change. A major conflict today between Western and Islamic societies is precisely this change in perceptions of what is morally correct; Islamic societies have not joined this social revolution.

The secularizing of the societies of the West has taken place rapidly. England is now sometimes referred to as a post-Christian country, and Scandinavian nations appear even more secularized. National churches have been either disestablished or face challenges in several European countries. The U.S. courts have not been especially hospitable to religious programs. Scientific developments, the pervasiveness of the media, the elimination of provincial isolation, urbanization, and rapid medical advances have all been factors in this secularization. Rising living standards of people in most Westernized countries have focused attention on what this life has to offer rather than on promises of some life to come.

A number of atheistic post-World War II movements have attracted the attention and frequently the loyalties of intellectuals. French existentialism, at least in its atheistic form, has pessimistically taught that it is ridiculous to believe a wise and benevolent God presides over this chaos. Consider the devastation of Europe in two wars of the 20th century—the first a long assault that wiped out a considerable part of the male population of Europe and the second a civilian bloodbath beyond all previous imagining. All concepts of moral and spiritual progress seemed demolished by these events. Have they served to confirm Malthusian and Darwinian precepts? How can Jewish philosophical minds find it possible to believe in a covenant God, or any benevolent Deity, after the Holocaust that almost destroyed Jewish life in Europe? Existentialism concludes that though the universe itself may be "absurd," to use the term favored by Albert Camus, the human mind attempts to impose meaning and must work out its own path.

As the 20th century progressed, traditional religion in the United States had much competition from nonreligious sources. Among the substitutes for religion have been nationalism, communism, psychoanalysis, and ecology. The cult of personality, where public figures, both loved and hated, are celebrated in the popular media, exalts the individual. Elvis Presley and Michael Jackson are only two of the several media personalities around whom spiritual cults have developed. Some music concerts have given the appearance of orgiastic religious rites.

Openly atheistic governments held sway in many parts of the world during the 20th century. While the Soviet constitution offered freedom of religion along with freedom of antireligious propaganda, many restrictions were placed on religious observance. Splendid historic churches were closed or became museums of the history of religion and atheism, detailing in lurid dioramas the atrocities committed throughout history in the name of religion. Party loyalists and often people important in their professions could not be church members, although some appear to have been secretly observant. Only a few active churches were allowed to remain open in large cities, even though they were filled with people during ceremonies. Synagogues and mosques were also closed, their clergy and teachers persecuted and sometimes killed. Only when the tourist trade became economically important was there an effort throughout the old Soviet Union to restore the most famous churches, synagogues, and mosques.

The 20th century brought about an extraordinary meeting of cultures like no time before in history. When peoples and cultures collide, ideas and beliefs are exchanged, with results both obvious and subtle. Missionaries, many of them from the Catholic and Protestant intelligentsia, served in Christian missions in India, China, and Japan. Some became scholars of comparative religion. They realized that religious systems—venerable, valuable, and filled with poetry and spiritual insight—already existed in these countries. This wisdom could be built upon rather than demolished. Many missionaries learned ancient languages, translated Asian holy books into European tongues, and, in a few cases, even assisted in the revival of the very religions they had been sent to replace.

A blending of religious ideas may be fruitful, but the increased marketplace of religious ideas, where many voices are contending, cannot help but have some effect on the skeptical mind. Are all religious ideas culturally conditioned? Are all major religions about equally valid or invalid? These are questions that must necessarily arise in a pluralistic society. Blaise Pascal, the 17th-century French philosopher and mathematician, proposed what became known as "the safe wager." If Christianity turned out to be false, he suggested, then the Christian would have lost no more than the atheist.

But if it turned out to be true, then the atheist would have considerable explaining to do in the next life. Any gambler, he believed, would have no hesitation in accepting the safe wager of faith. Now, with so many religious ideas to choose from, and Hindu and Islamic families possibly living next door, how can any American find Pascal's choice so simple?

Novel ways of practicing religion have also presented themselves. Televangelism is a striking phenomenon that has troubled both established churches and secular society alike. It was in part this development, coupled with biblical literalism, that led Langdon Gilkey to his impassioned disavowal of intelligent design. Televangelism has enabled people to worship in their living rooms, request healing of spiritual and physical sickness by touching a television set, and send their tithes to well-combed, wealthy televangelists. Billy Graham, a sincere evangelist who has not been one of the money-grubbers, has used television conscientiously, addressing more people in one internationally televised sermon than all prophets and preachers of the past during entire ministries.

But even televangelism may ultimately be supplanted by the virtual church, as the World Wide Web is quickly embraced by evangelists. Religious chat lines, support groups, deprogramming groups, as well as virtual religious communities are flourishing. Imagine a church with a million members but no visible community, no concrete sacraments, no fleshly contact with other human beings. Even the face of the pastor may be hidden.

Religions have given meaning to existence, have provided answers to the basic questions in life, and have given consolation in times of suffering. They have provided a sense of security amid the flux and transitions of life, an anchor amid changes and disruptions. Religions have also provided poetry, mystery, and color to otherwise drab lives. Their ceremonies have marked the important events in human lives: birth, marriage, and death. Religious organizations have created, perhaps more capably than any other institutions, brotherhoods, sisterhoods, communities with common aspirations. In their teachings and sacraments, they have also forged a link between the living and the dead. Pilgrimages to revered places, prayers for the dead, and ancestor worship have reinforced the belief that the faithful are part of a communion across time and space. Churches have often bestowed full social lives on the lonely. The study of holy books, theologies, and church histories has provided intellectual stimulation to others, promoting both religious and secular learning.

Though good conduct may be taught in ethical cultural societies, churches, synagogues, and mosques are much more effective in that they remind people that their actions take place in the presence of the Deity. People will certainly behave better toward their neighbors if they are convinced that all are brothers and sisters and that God is looking over their shoulders. Secular rites never have the force of religious ones. For example, in communist Russia the ceremonies in state-operated marriage parlors never competed successfully with the elaborate marriage ritual of the Orthodox Church.

Of course, as Richard Dawkins and others are constantly reminding their readers, religions, all of them, are tainted with the pathology that manifests itself in every human endeavor. The atrocities committed in the name of religion have been numerous, and they are not all calamities of the past. Religions still sometimes retard knowledge, holding people in a benighted ignorance. Members of the clergy sometimes exploit people and engage in despicable acts. Religion has often maintained its hold over minds through fear rather than promise. Some religions teach reprehensible doctrines, enslave people, and teach them to be content with their lot rather than work to improve it.

A RELIGION OF SCIENCE

Can science ever replace religion? Can it meet human spiritual and emotional needs, just as it has done a meritorious job of meeting physical needs? What do people seek from religion that makes it still so popular throughout the world even in the 21st century, despite all the attacks upon it? How will the questions to which religion has provided at least partial answers be addressed in an age that enthrones science? Science now has high prestige. We know that some religions have even appropriated the word *science* in the way they identify themselves. We have Christian Science, Scientology, Science of the Soul, and other such religious organizations, though their critics are adamant in claiming that they have misused the word.

There have been a few ambitious attempts to construct religions based on the prestige and numinous quality of science itself. The Institute of Religion in an Age of Science was founded in 1954, the work chiefly of Ralph Burhoe and Harlow Shapley. It was Harvard University–based, influenced by Moral Rearmament, the Unitarian Church, and the writings of Julian Huxley. Filled with cosmological speculation and a faith in the evolutionary process operating throughout a universe populated by life, it sought to establish a Coming Great Church for future generations, which would employ science as the ultimate way of exploring religious questions. The founders believed that the wholeness of society depended upon such a workable faith.[6]

Throughout the 1950s, Burhoe and Shapley talked of lifting wisdom selectively from ancient scriptures and compiling a modern Bible that would defer to the accumulation of modern scientific knowledge. They had long concluded that some form of religion was inevitable and necessary. Had not anthropologists found spiritual expression in all societies, both primitive and complex? The program of Burhoe and Shapley included a plan for scientists, in dialogue with theologians, to teach "scientific theology" to university students. With this in mind, they founded in 1966 a professional journal of science and religion, *Zygon,* published by the University of Chicago Press, with Burhoe as editor.

Despite their noble aspirations, the founders of Coming Great Church did not have much success. Many scientists and theologians were still committed to the view that their disciplines occupied separate realms and were best left to attend to their own concerns only. Burhoe and Shapley, rationalists themselves, marveled that Unitarianism and other liberal movements had been unable to "sweep through the world." They also found, as the world moved into the 1960s, with the revival of primitive cults and countercultures, that irrationalism was more powerful than their sensible goals.

Probably the reason for the lack of popular success of Coming Great Church was best expressed by Methodist bishop Francis Gerald Ensley, who observed: "An artificial religion is about as consequential as Esperanto." He noted that any living religious faith had to be "continuous with a historical tradition and be embodied in a cult and a community."[7]

RELIGION IN THE FUTURE

Looking out upon a postmodern age, it is tempting to speculate on what may be in the future for world religion. Most prophecies, no matter how well informed the prophets may be, turn out to be wrong. No doubt, there will be many events we cannot anticipate at the beginning of the 21st century. Will there be a revival of tribalism around the world, or will ethnic groups and nationalities lose their distinguishing features and traditions as human society becomes more and more global? Will religions continue to be entangled in national and regional conflicts, or will secular forces intervene? Will terrorist movements operating in the name of religions become more fierce and feared, with weapons of mass destruction at their disposal, or will religious enthusiasm be channeled into spirituality, artistic expression, and productive social causes? Will there be a resurgence of clerical influence or will the individual take charge of personal spiritual welfare? Will the influence of popes, rabbis, lamas, priests, and muftis increase or decline?

Will the great religions of the world accommodate themselves to science and the modern world, or will there be increasing revivals of fundamentalist movements rejecting science in Judaism, Christianity, Islam, and the other religions? Will existing religions continue, retaining or extending their influence? Will they undergo reformations, possibly including radical changes? Will ancient, "infallible" scriptures be deconstructed and adapted? Will priesthoods be expanded to all genders and peoples? Will concepts of what is morally permissible change in the direction of ever more freedom for personal inclinations? Will challenged practices such as abortion and euthanasia, along with changing family styles, be accepted in comfortably accommodating religious communities? Or will the pendulum swing, with the discovery that freedom often brings chaos? Will dominant spiritual traditions make common cause to achieve mutually desired goals, or will competition between them increase on the mission fields and within nations?

Will a religion or cult now considered minor become the great faith of a new age? Will Mormonism, or Bahai, or some now-obscure movement become the world faith of tomorrow? Will there be a blending of religious ideas from East and West, adopting the most attractive features of each, while eliminating the crude superstitions and outmoded practices that no longer seem to work? Will a great super-religion emerge? Will a science fiction religion, somewhat like Scientology but with broader appeal, seize the human imagination? Will a space-age religion unlike any now known come into being to satisfy the demands of a new era? In an age of personality journalism, is it unreasonable to ask if a new messiah, prophet, Muhammad, Christ, or Buddha will come forth? Will such a person start a revolutionary spiritual movement that will immediately attract people of all races and nations, uniting them in a global village? And will the mission of such a new leader be swiftly facilitated by the media which even now reaches almost all corners of the globe?

And finally, how will all this influence education? These and other questions remain to be answered. The questions are known, the answers are not. The task of education, if it is to be effective in the 21st century, is immense.

NOTES

1. C. P. Snow, "The Two Cultures," in *The Scientist and the Humanist,* eds, George Levine and Owen Thomas (New York: W.W. Norton, 1963), 1–6.

2. W. T. Stace, "Man against Darkness," in *The Search for the Meaning of Life,* ed. Robert F. Davidson (New York: Holt, Rinehart and Winston, 1965), 210–217.

3. Ibid., 213.

4. Ibid., 216.

5. Ibid., 217.

6. James Gilbert, *Redeeming Culture: American Religion in an Age of Science* (Chicago: University of Chicago Press, 1997), 273–295.

7. Ibid., 291.

Appendix

PERSONALITIES OF THE EVOLUTION-CREATION DEBATE

Controversies are in part always driven by the personalities of the people who dominate them. A number of unusual, highly creative individuals have put their stamp on the evolution-creation debate. Some are among the movers and shakers of the last centuries. Some are world famous, while others, equally interesting, are less well known. The list that follows is alphabetical and does not attempt to rank in any order of importance.

Aristotle (384–322 B.C.), Greek philosopher often called the father of Western science. A student of Plato and tutor to Alexander the Great, Aristotle believed in empirical research and classified living beings in a graded scale, which, with later adaptations, would become known as the Great Chain of Being.

Saint Augustine (A.D. 354–430), one of the Latin fathers of the church, associated with the cultural seat of ancient Alexandria. In his youth, he resisted his mother's Christianity, more attracted to the paganism of his father. Later, after a conversion in a garden in Milan, he became Bishop of Hippo in North Africa. A prolific writer and one of Christianity's greatest theologians, Augustine believed that true science and true religion would never conflict. He also believed that the words of Holy Scripture must be understood as accommodations to the understanding of the people to whom they had been first revealed.

William Bateson (1861–1926), British geneticist associated with Cambridge University. Bateson coined the term *genetics* to describe the study of biological inheritance. He did more than anyone else to make known the findings of Gregor Mendel after his work in heredity was rediscovered in 1900. From a distinguished family, Bateson was, in turn, the father of the noted anthropologist, Gregory Bateson.

Michael J. Behe (1952–), professor of biological sciences at Lehigh University and leading intelligent design champion. Behe, a biochemist, is best known for his argument for irreducible complexity, which he believes supports ID. He is a practicing Roman Catholic.

Friedrich von Bernhardi (1849–1930), Prussian general and military historian and theorist. He was a major influence in Germany prior to World War I. Building upon his own interpretation of Darwinian principles, he advocated ruthless aggression, disregard of treaties, and war as a "divine business." In his book *Germany and the Next War* (1911), he called war "a biological necessity," a manifestation of the struggle for existence that rests on natural law.

Carrie Buck (1904–1981), plaintiff in the *Buck v. Bell* case before the U.S. Supreme Court. Early in life she was declared feeble-minded and forced to undergo compulsory sterilization, in accord with the eugenics program of the state of Virginia. She was later married for 25 years to William Eagle and revealed to be a woman of normal intelligence. One of the sorrows of her life was the inability to bear children during her long, successful marriage.

Brandon Carter (1942–), Australian theoretical physicist known for his work on the properties of black holes. He appears to have been the first scientist to clearly identify and name the anthropic principle, the extraordinary hospitality of the earth to the development of human life.

Francis S. Collins (1950–), geneticist widely acknowledged as one of the most significant scientists of the 20th and 21st centuries. His discoveries of disease genes and his leadership in the Human Genome Project are his best known accomplishments. He was appointed in 2009 director of the National Institutes of Health, the year he founded the Biologos Foundation to give a public voice to those who find science and religion mutually enhancing. In 2006, he published a popular book, *The Language of God: A Scientist Presents Evidence for Belief,* in which he identifies scientific discoveries as opportunities for worship. Although he is a devout Christian who has debated the atheistic biologist Richard Dawkins, Collins rejects the creationist and intelligent design movements.

Ann Hart Coulter (1961–), American social and political commentator, television personality, and best-selling humorist. Coulter admittedly likes to "stir up the pot" and is known for her highly controversial assertions. In her 2006 book *Godless: The Church of Liberalism,* she characterized evolution as "bogus science." She believes that the Left, her major enemy, has an obsession with Darwinism as a replacement for religion. She identifies herself as a Christian, usually attending the Presbyterian Church.

Clarence Seward Darrow (1857–1938), American lawyer, famous for representing the defendants in the Leopold and Loeb trial in Chicago and the Scopes trial in Dayton, Tennessee. Though he was born in Ohio, Darrow's home base became Chicago, where he was a vocal civil libertarian and religious agnostic.

Charles Darwin (1809–1882), British gentleman-scientist who is arguably the most influential person of the modern age. It was Darwin's presentation of the origin of the species that changed scientific thinking all over the world and challenged traditional religious views.

Erasmus Darwin (1731–1802), English physician, natural philosopher, and grandfather of Charles Darwin and Sir Francis Galton. A member of the gifted Darwin-Wedgwood family, he was an eccentric, lively personality. He expressed his own primitive evolutionary views in a long, dull poem.

Charles Davenport (1866–1944), Harvard University biologist and leader in the eugenics movement in the United States. This movement resulted in the sterilization of over 60,000 people and is believed later to have provided ideological support for the German Holocaust. Davenport was a strong opponent of miscegenation and encouraged state laws forbidding "race-crossing." He was associated with influential German journals and maintained these connections in Nazi Germany, even during World War II. Today his work is widely dismissed as racist, elitist, and unscientific.

Paul Davies (1946–), British cosmologist and astrobiologist, later a professor at Arizona State University and director of Beyond, the Center for Fundamental Concepts of Science. In 2005, he became the chair of the SETI project, which searches for signs of intelligent life beyond earth.

Richard Dawkins (1941–), British evolutionary biologist and skilled popular science writer, associated with Oxford University. An aggressive atheist with a distaste for all religion, he is best known for his book *The Blind Watchmaker* (1986), a detailed argument against the watchmaker analogy used in natural theology to affirm the existence of a Creator. Dawkins has been widely referred to as "Darwin's Rottweiler." He believes that Darwin has now made atheism inevitable for thinking people, while religious faith remains a troublesome delusion. In *The Selfish Gene* (1976), he introduced his concept of the meme, a cultural equivalent of a gene, used to explain the spread of ideas and cultural phenomena.

William A. Dembski (1960–), mathematician, philosopher, and a leading U.S. proponent of intelligent design. Although he has been associated with Protestant seminaries, Dembski is an Eastern Orthodox Christian.

Daniel Dennett (1942–), American philosopher of science and evolutionary biology, associated with Tufts University. He is a forceful atheist who calls for all religions to be taught in the classroom so that students can understand that religion is just another natural phenomenon. He has argued with Stephen Jay Gould, who describes Dennett's thought as "Darwinian fundamentalism." In his 2006 book, *Breaking the Spell: Religion as a Natural Phenomenon*, Dennett attempts to discover the origins of religion and explain its hold over human lives. He contends that the protective wall of mystery that surrounds religion must be removed so that its nature and function may be more fully understood.

Michael Denton (1943–), British-Australian biochemist. In his book, *Evolution: A Theory in Crisis* (1985), Denton challenged neo-Darwinism and presented evidence for supernatural design in nature. His book was highly influential at the beginning of the intelligent design movement. However, his views later were modified, and with his second book, *Nature's Destiny* (1998), he argued for a patterned evolutionary unfolding of life.

Theodosius Dobzhansky (1900–1975), Ukrainian-born geneticist and evolutionary biologist, a central figure in the modern evolution synthesis. His most important work was completed in the United States. Throughout his life, he remained a communicant in the Eastern Orthodox Church.

Niles Eldredge (1943–), American paleontologist, scientific writer, and curator of the Department of Invertebrates at the American Museum of Natural History. With colleague Stephen Jay Gould, he developed the evolutionary theory of punctuated equilibria to explain the extraordinary variety of fossils known as the Cambrian explosion.

Sir Francis Galton (1822–1911), British scientist, Renaissance man, and cousin of Charles Darwin. He applied statistical methods to the study of human diversity and the inheritance of intelligence, developing methods that attempted to measure human intelligence. He also made valuable contributions to forensic science but is best known as an early advocate of the pseudo-science that he named eugenics. Believing that humans should be selectively bred much as farm animals are, he advocated early marriage for people of high rank and achievement, to be encouraged by monetary incentives.

Langdon Brown Gilkey (1919–2004), educator and American Protestant theologian. Gilkey taught at Vassar College, Vanderbilt Divinity School, and the University of Chicago Divinity School, among other institutions. He was interested in the interactions between science and religion, opposed Protestant fundamentalism, and was an expert witness for the prosecution in the *McLean v. Arkansas* trial. He believed that a "rough parity" existed among world religions and found Buddhism especially inspiring.

Owen Gingerich (1930–), former research professor of astronomy and the history of science at Harvard University and senior astronomer emeritus at Smithsonian Astrophysical Observatory. A flamboyant and popular professor, Gingerich is known for his dramatic classroom antics. He was born into a Mennonite family in Kansas. Though still a theist, he believes intelligent design supporters inadequately explain the distribution of species on earth and the significance of DNA coding. Evolution, he contends, offers more satisfying scientific explanations.

Stephen Jay Gould (1941–2002), American paleontologist, evolutionary biologist, and popular science writer, one of the best known scientists of the last half of the 20th century. His chief contribution to science was the punctuated equilibria theory of evolution, presented with Niles Eldredge in 1972. Gould and Eldredge taught that evolution is marked by long periods of stability, interrupted by periods of rapidly branching development. This contrasts with the more widely held theory that evolutionary change is constant.

Asa Gray (1810–1888), renowned American botanist of the 19th century. He developed the accepted classification of North American plants, published extensively, and created the botany department at Harvard University, where he was professor. He arranged publication of the first American edition of Darwin's *On the Origin of Species* and communicated frequently with Darwin. A devout Presbyterian, Gray obtained an admission from Darwin that evolutionary theories were not antithetical to religious faith. Throughout his life, Gray contended that orthodox Protestant Christianity could be reconciled with evolution.

Ernst Haeckel (1834–1919), German biologist who promoted Darwin's work in Germany and propounded the theory of recapitulation, claiming that an individual's fetal development parallels the evolutionary development of the species. Haeckel's drawings of embryos, which still appear in numerous textbooks, are now considered bogus. He has also been accused of falsifying much of the data that made him famous.

Fred Hoyle (1915–2001), English astronomer, science fiction writer, and one of the more interesting and independent thinkers of the 20th century. His chief research was conducted at Cambridge University, where he served as director of the Institute of Astronomy. He is best known for his contribution to the theory of stellar nucleosynthesis, according to which chemical elements were synthesized from the primordial hydrogen and helium found in stars. Hoyle aroused controversy with his unorthodox cosmological views. Although he coined the term Big Bang, he rejected that cosmological theory. An atheist in early life, Hoyle moved toward belief in a Higher Intelligence through his scientific work and a theory of origins he called panspermia, developed with his student and colleague Chandra Wickramasinghe. In the Hoyle-Wickramasinghe book *Evolution from Space* (1982), they proposed that evolution on earth took place through the influx of viruses brought to the planet by comets. Hoyle famously compared the chance emergence of even the simplest life, according to conventional theories of evolution, to a tornado assembling a Boeing 747 by sweeping through a junkyard. In other writings, he further rejected features of Darwinism and argued, against majority scientific opinion, for a steady state universe.

Julian Huxley (1887–1975), English evolutionary biologist, and grandson of Thomas Henry Huxley. He led an attack on Trofim Lysenko's theories, so influential in the Soviet Union, and brought the work of Pierre Teilhard de Chardin to the English-speaking public. He described himself as an atheist.

Thomas Henry Huxley (1825–1895), English biologist, widely known as "Darwin's bulldog" for his enthusiastic advocacy of Darwin's theory of evolution. His 1860 debate with Bishop Samuel Wilberforce was an important event in the history of the reception of evolutionary theory. Huxley was a strong advocate of scientific education in Great Britain and originated the term *agnostic* to describe his own religious views. Although he was largely self-taught, he was widely recognized as the leading comparative anatomist of the last part of the 19th century.

Pope John Paul II (1920–2005), Roman Catholic pontiff, born Karol Jozef Wojtyla in Poland. He found evolution compatible with Christianity, though each human soul, he taught, is individually created and implanted by God. In an important speech to the Pontifical Academy of Sciences in 1996, John Paul II recognized evolutionary theories to be factual, even though they contradicted a literal reading of Genesis. He taught that there is a right and wrong way of reading both the scriptures and science. This pope is also admired for his role in ending European communism and his peaceful overtures to Eastern Orthodox and Protestant Christians. He was the first pope to visit a synagogue in Rome, where he referred to Jews as "our elder brothers in the faith."

Phillip E. Johnson (1940–), professor emeritus of law from the University of California, Berkeley, and one of America's most distinguished legal scholars. Early in his career, Johnson served as law clerk for U.S. Supreme Court Chief Justice Earl Warren. Johnson, who experienced a religious conversion in midlife and became a born-again Christian, is one of the most important critics of Darwinian evolution and a founder of the intelligent design movement. He contends that the evidence for Darwinian evolution is so thin that it could not stand up in a legal court. He is the chief formulator of the Wedge Document which seeks to spread ID thinking and outlines strategy. His books, including *Darwin on Trial,* have sold widely.

Paul Kammerer (1880–1926), Austrian biologist and skilled musician, one of the most glamorous and tragic figures in the history of science. In his attempt to prove the Lamarckian theory of inheritance, he experimented with midwife toads. He believed that in his breeding of these toads, and the emergence of black pads on their feet, he had established the existence of hereditary acquired traits. He was later accused, perhaps falsely, of injecting his specimens with black ink in order to support his theory. He may have been undermined by a Nazi sympathizer at the University of Vienna, though the exact details of his downfall are mysterious. Though he was offered an important position in Russia, he committed suicide in disgrace.

Hans Küng (1928–), Swiss Roman Catholic priest, liberal theologian, and university professor. Although no longer allowed to designate himself a Catholic theologian because of his unorthodox ideas, Küng is still an influential teacher who has written meaningfully on science and religion.

Jean-Baptiste Lamarck (1744–1829), French naturalist and early evolutionary theorist. He appears to have been the first person to designate his particular field of study as botany. Lamarck provided the first systematic theory of evolution, proposing an alchemical force driving organisms to ever higher levels of complexity along with an environmental force that adapts them to their environments. He also taught that acquired characteristics may be inherited by future generations. His most famous example was the giraffe, believed to have evolved its long neck in order to reach food on tall tree branches. This theory of inheritance, which Darwin accepted at least in part, is now rejected.

George-Louis Leclerc, Comte de Buffon (1707–1788), French naturalist, mathematician, cosmologist, and encyclopedic author. His prolific writings encompassed everything known about that natural world in his time. Darwin identified him as the first author of modern times to contemplate evolution in a scientific spirit, though Leclerc eventually concluded that species were immutable. He did, however, note the similarities between humans and apes and at least entertained the possibility of a common ancestor.

Charles Lyell (1797–1875), British lawyer and geologist, best known for his presentation of "uniformitarianism." In *Principles of Geology* (published in 11 editions, beginning in 1830), his most famous work, he proposed an ancient earth. Lyell was personally a religious man, even as he rejected biblical chronology. His influence on Charles Darwin was strong.

Trofim Lysenko (1898–1976), Ukrainian biologist and agronomist, largely self-taught. Stalin made him director of Soviet biology and made his theories official dogma. Lysenko rejected Mendelian genetics, believed in the inheritance of acquired characteristics, and adopted questionable hybridization theories with disastrous results for Soviet agriculture. Stalin's enforcement of his theories almost destroyed Russian genetic research, which had once been world-renowned. Today the word *Lysenkoism* refers to fraudulent scientific theories enforced by a government.

Thomas Robert Malthus (1766–1834), British clergyman whose work in political economy and demographics was a major influence on Charles Darwin. Malthus felt that hopes for progress toward a utopian society, so popular with thinkers of his time, were challenged by the dangers of population growth. Because unchecked population growth would eventually swamp the world's resources, a balance in population is brought about by epidemics, pestilence, plague, and wars. He was critical of British poor laws, believing that God allowed excessive population growth for moral purpose, to teach the virtues of hard work and sound behavior. He also taught that evil exists in the world to spur activity. With the abolition of poor laws, be believed that dire need should be addressed by private charity. Though Malthus was himself the youngest of eight children, he limited his own brood to only three.

Henry Louis Mencken (1880–1956), American journalist, known as the Sage of Baltimore. Mencken was an acerbic critic of U.S. culture and the author of a famous book, *The American Language*. His reporting of the Scopes trial had much to do with the way that event is remembered in history. Always quotable, he referred to the American South as "the Bible and Syphilis Belt." He also held racist and anti-Semitic views.

Gregor Johann Mendel (1822–1884), Austrian monk and one of the greatest scientists in history. He entered a monastery in 1843, taught in a local secondary school, and later became the abbot of his order. Through his experiments with pea plants in the monastery garden, he discovered the basic laws of heredity. It was his work, largely ignored at the time of its first publication, that later helped vindicate Darwinian theories and led to the neo-Darwinian synthesis.

Stephen C. Meyer (1958–), widely acknowledged as cofounder of the intelligent design movement and senior fellow of the Discovery Institute, Meyer is a historian of science and a geophysicist.

Kenneth R. Miller (1948–), biology professor at Brown University and a practicing Roman Catholic. His book *Finding Darwin's God* contends that evolution is not incompatible with Christianity. Yet he is a strong critic of creationism and intelligent design and was an expert witness for the prosecution in the *Kitzmiller v. Dover Area School District* case.

Robert A. Millikan (1868–1953), American teacher and scientist, best known for discoveries in electricity, optics, and molecular physics. In 1923, he was awarded the Nobel Prize in physics. His religious commitment was expressed in numerous writings; of particular note is *Evolution in Science and Religion* (1927).

Irwin Moon (1907–1981), American founder of the Moody Institute of Science and creator of a series of highly popular "sermons from science" films. Moon believed science and conservative religion could be reconciled.

Henry M. Morris (1918–2006), American young earth creationist and a founder of the Creation Research Society and the Institute for Creation Research. He is sometimes referred to as the father of modern creation science. A Texas evangelical Christian, he was a hydraulic engineer and taught in several important universities. As coauthor of *The Genesis Flood* (1961), perhaps the most influential text for young earth creationists, his influence extended throughout the North America.

Origen of Alexandria (ca. A.D. 185–ca. 254), one of the most important of the early Greek fathers of the church. In his many writings, which synthesized Greek philosophy with Christian doctrine, he taught that there are three levels of meaning in sacred scripture. The bodily level, the bare letter of the text, meets the needs of simple folk. The psychic level, which provides deeper understanding, assists in the soul's progress toward perfection. The spiritual level of interpretation is the highest in that it treats the "unspeakable mysteries" that make humans partakers of the wisdom of the Holy Spirit.

William Jay Overton (1939–1987), judge of the United States District Court for the Eastern District of Arkansas, known for his ruling on Act 590 in *McLean v. Arkansas.* In his ruling against the state, Judge Overton found that creationism is not a scientific theory because it is dogmatically absolutist rather than subject to revision according to proper scientific procedure. His decision has been both admired and reviled.

William Paley (1743–1805), British philosopher and Christian apologist. As a justice of the peace and an archdeacon in the Church of England, he was a forceful opponent of the slave trade. He is best remembered as a defender of Christianity in his books *View of the Evidences of Christianity* (1794) and *Natural Theology; or, Evidences of the Existence and Attributes of the Deity* (1802), which were once required reading in British universities and are still valued by religious fundamentalists. He famously compared God to a master watchmaker. The young Charles Darwin faithfully studied these books in university.

Philo of Alexandria (20 B.C.–A.D. 50), Hellenistic Jewish theologian and philosopher, a native and inhabitant of the ancient cultural center of Alexandria, Egypt. Philo's thought is believed to have exerted a strong influence on early Christian theology. Certainly, his methods of biblical interpretation were widely adopted by early Christian thinkers. Although the Hebrew Bible had not been definitively canonized in his time, Philo taught that the Torah (First Five Books of Moses) were infallible revealed truth. Still, he believed that there are two unequal ways of interpreting scripture, the literal and the allegorical. According to his rules of interpretation, certain passages of scripture must not be taken literally, because, on their surface, they are unworthy of God and even senseless and contradictory. Philo found the allegorical approach the more authentic, although open only to the initiated.

Pope Pius XI (1793–1878), born Giovanni Maria Mastai-Ferretti, the longest-reigning pontiff in the history of the Roman Catholic Church. He convened the First Vatican Council in 1869, which defined papal infallibility under limited circumstances. He was skeptical of evolutionary theories, regarding them as a challenge to the faith.

John Polkinghorne (1930–), British particle physicist, Anglican priest and theologian, winner of the Templeton Prize in 2002. For 25 years, Polkinghorne worked on theories of elementary particles and was instrumental in the discovery of the quark. For more than ten years, he was a professor of mathematical physics at Cambridge University, before resigning his professorial chair to study for the ministry. He first became curate in a large working-class parish in Bristol before returning to Cambridge as dean of Trinity Hall Chapel. Polkinghorne believes that science and religion are both avenues to divine knowledge, the universe is intelligible, and human free will exists. He remains one of the most forceful scientific proponents of Christianity.

William G. Pollard (1911–1989), physicist and Episcopal priest. Sometimes referred to as "the atomic deacon," he was executive director of the Oak Ridge Institute of Nuclear Studies. In discussing free will, Pollard frequently evoked quantum indeterminacy and chaos theory. His book *The Frontiers of Science and Faith: Examining Questions from the Big Bang to the End of the Universe* (2002) is considered a basic work in the science-theology debate.

George McCready Price (1879–1963), Canadian-born Seventh-Day Adventist educator, one of the central figures in the creationist movement. His most influential publication was *The New Geology* (1923), a 726-page college textbook refuting Darwin's theory of evolution. Price attacked the dating techniques of evolutionary scientists, contending that all fossils were laid down at the same time during Noah's flood. Williams Jennings Bryan appealed to Price's work during the Scopes trial in Dayton, Tennessee. Price's influence on Henry M. Morris and John Whitcomb in the 1960s is reflected in their book *The Genesis Flood*. Price believed that Noah's flood was worldwide, and he rejected both the day-age and gap creation theories.

Michael Ruse (1940–), English philosopher, important for his writing on the creation-evolution controversy. He was an expert witness for the prosecution in the *McLean v. Arkansas* case and has recorded his impressions. He believes the Christian religion can be reconciled with evolutionary theory but that intelligent design arguments are wrongheaded.

Bertrand Russell, 3rd Earl Russell (1872–1970), English philosopher and social critic, one of the most influential thinkers of the 20th century. His most pertinent contributions to the religious-science controversies were *What I Believe* (1925) and *Why I Am Not a Christian* (1927).

John Thomas Scopes (1900–1970), American geologist and engineer who was the defendant in the 1925 "Monkey Trial" in Dayton, Tennessee. Scopes later worked in South America and Louisiana. Upon his marriage, he became a nominal Roman Catholic. In his memoir, *Center of the Storm* (1967), he corrected many misconceptions surrounding the Dayton event.

Gerald Schroeder (no dates available), American Israeli scientist, speaker, and writer. A 1965 PhD from the Massachusetts Institute of Technology in nuclear physics and planetary sciences, Schroeder joined the Weizmann Institute of Science in Israel in 1971 and has been associated with Hebrew University in Jerusalem. In his work, Schroeder reconciles modern science with the Bible, the Talmud, and the Jewish mystical tradition. He reminds his audience that one day is as a thousand years in the mind of God, and this should be kept in mind in interpreting the Genesis chronology of Creation. He does not reject as much of the literal meaning of the Hebrew Bible as do many scholars, suggesting that Neanderthals and Cro-Magnons may have existed before God breathed spirituality into Adam, the first real human, about 6,000 years ago.

T. O. Shanavas (dates not available), pediatrician who immigrated to the United States from India in 1970. Shanavas is active in the Islam Center of Greater Toledo, Ohio, and the Islamic Research Foundation in Louisville, Kentucky. A current resident of Michigan, he writes on the relationship of religion and science from the Islamic perspective. In reviewing the learning of the Islamic Golden Age, he finds the insights of modern science in harmony with the Qur'an and anticipated by the early philosophers of Islam. His book *Creation And/Or Evolution: An Islamic Perspective* (2005) is basic to any discussion of the Abrahamic religions and modern science.

George Bernard Shaw (1856–1950), Anglo-Irish playwright with ironic views on almost all subjects. He won the Nobel Prize for literature in 1925 and regarded himself as greater than Shakespeare. He was a socialist, a member of the Fabian Society, and an iconoclast who did not accept the dogmas of any religious organization. Equally skeptical in scientific matters, he accepted what he called creative evolution.

George Gaylord Simpson (1902–1984), American paleontologist associated with Columbia University, possibly the most influential figure in his science in the 20th century. He was a major participant in the neo-Darwinian synthesis, anticipating such concepts as punctuated equilibria. Although he did comment on the general absence of transitional forms of plants, animals, and humans in the fossil record, he was a strong proponent of evolution, teaching that humans are the result of a purposeless natural process that does not in any way favor human development.

Charles Percy Snow (1905–1980), British physicist and novelist who held a number of government positions. He is widely known for his lectures and writings on "the two cultures" and his conviction that the humanities and sciences need to dialogue with each other.

Herbert Spencer (1820–1903), English philosopher and sociological theorist. He coined the phrase "survival of the fittest" in his *Principles of Biology* (1864). He accepted Lamarckian heredity and had problems with Darwin's principle of natural selection. A romantic, he was strongly influenced by the poet Samuel Taylor Coleridge. Unlike Darwin, Spencer believed that evolution had a clear direction, moving toward an end state, which he designated as equilibrium. In this respect, he may have been a precursor of Teilhard de Chardin.

Walter Terence Stace (1886–1967), British civil servant, educator, and philosopher. Stace served in the Ceylon Civil Service and held positions in the Ceylonese government. He accepted an endowed chair in philosophy at Princeton University in 1935 and served as president of the American Philosophical Association in 1949 and 1950. Always interested in religion, his book *Mysticism and Philosophy* (1960) is noted for its balance and clarity. He wrote of the difficulty of belief in an age of science, yet felt asking for proofs of God's existence was the equivalent of asking for proofs of the existence of beauty.

William Graham Sumner (1840–1910), Yale University sociologist who taught laissez-faire economics, which he justified by appeals to Darwin's laws of evolution. Although he did not advocate extermination of what he regarded as the unfit classes, he did believe certain individuals were detrimental to society. Eugenicists found his views helpful. He was also an ordained Episcopal priest.

Pierre Teilhard de Chardin (1881–1955), French paleontologist, philosopher, and Jesuit priest. Teilhard took part in the discovery of Peking Man and emerged as one of the leading paleontologists of the first half of the 20th century. He is also a central figure in any study of the relationship between science and religion. Although he was often in conflict with his religious order because of his understanding of original sin, his work finally received papal recognition in 2009. He believed in progressive evolution, or ontogenesis, with a gradual unfolding of the material cosmos, leading toward an Omega Point in the future, when all creation would achieve its consummation in the consciousness of God. Many of Teilhard's statements seem to suggest a near-Eastern form of religious mysticism.

Charles Thaxton (1939–), American creationist, author, and fellow of the Discovery Institute's Center for Science and Culture. Thaxton is a physical chemist with postdoctoral work in the history of science at Harvard University and molecular biology at Brandeis University. His best-known publication is his edition of the popular school textbook, *Of Pandas and People* (1989). In his speaking and writing, Thaxton attempts to stay within the empirical domain, not bringing God directly into the discussion. Still, he asserts that special creation was the work of a Designer.

James Ussher (1581–1656), Anglican bishop of Armagh and one of the great theological scholars of his day. From his studies of biblical genealogy, he concluded that the world was created in 4004 B.C., a calculation widely accepted and often printed in the margins of Protestant Bibles. He also regarded the pope as antichrist, an idea not surprising for his time and place but considered indefensible today. Ussher's work in patristics is still regarded as germinal, and he was buried with honors in the St. Erasmus Chapel of Westminster Abbey.

John C. Whitcomb (1924–), American theologian and young earth creationist. His best-known publication (with Henry M. Morris) is *The Genesis Flood* (1961), a central writing in the creationist movement. After graduating from Princeton University with honors in ancient and European history, Whitcomb taught and lectured at conservative seminaries. As president of Whitcomb Ministries, he has been a popular speaker for Answers in Genesis. He is also an elder in the conservative Grace Brethren Churches. Although his writings have been criticized as scientifically and factually inaccurate, he remains one of the most influential creationists.

Ellen Gould White (1827–1915), founder of the Seventh-Day Adventist Church, one of the few religious organizations in the world originated by a woman. Her visions, which became the foundation of the unique Adventist doctrines, not only affirmed the biblical accounts of Creation as objective reality but also foretold the end of the ages. In addition to faith in the scriptures, she advocated good health habits, virtuous living, the education of youth, liberation of the oppressed, and compassion for all suffering.

Alfred North Whitehead (1861–1947), British mathematician and philosopher. Whitehead shared a family concern with theology; his father and uncles were vicars in the Church of England, and a brother became Bishop of Madras. Influenced by his Irish wife, Whitehead had leanings toward Roman Catholicism but never formally joined a church. After studying the work of Albert Einstein, he developed what has become known as process philosophy, of great influence on the school of process theologians. In his 1925 publication *Science and the Modern World,* Whitehead articulated his metaphysical views. In *Process and Reality* (1929), he defended theism, though not biblical religion. The universe is in a constant state of flux, Whitehead suggested, where God, too, is always growing and changing, unlike the static divinity of classical pagan and Christian philosophy.

Chandra Wickramasinghe (1939–), colleague of Fred Hoyle and professor of applied mathematics and astronomy at Cardiff University. A native of Sri Lanka, Wickramasinghe is chiefly active in the field of astrobiology, developing methods for detecting life in space. Always controversial and constantly creative, he propounded with Hoyle the theory known as panspermia. He believes that cosmic dust available in interstellar space and in comets originally seeded life on earth. Wickramasinghe has also suggested that some diseases might originate outside the earth.

Samuel Wilberforce, Bishop of Oxford (1845–1869), English cleric and member of a distinguished family. Wilberforce is best remembered as the man who debated Darwinism with Thomas Henry Huxley, "Darwin's bulldog." Although some Darwinists have painted Wilberforce as a pompous fool, many who attended the debate considered it a draw.

Selected Annotated Bibliography

In order to clearly understand the evolution, creation, and intelligent design controversies, it is helpful to know something of the legal, social, and religious background of the participants, as well as the changing attitudes of the American people. The following books, available in most well-equipped academic libraries, offer such a background as well as information on related issues that stimulate further investigation. There is probably no other issue that provides more insight into American habits, ideals, and attitudes, and which brings major cultural divides into sharper focus.

Aczel, Amir D. *The Jesuit and the Skull: Teilhard de Chardin, Evolution and the Search for Peking Man.* New York: Riverhead Books, 2007.
> Disregard the catchy title. This uneven book, though lacking a clear focus, summarizes basic information on human evolution, the mysterious disappearance of Peking Man's bones, and the life of Teilhard de Chardin. Some attention is given to Teilhard's complicated relationship with Lucile Swan and the priest's last years.

Ayala, Francisco J. *Darwin and Intelligent Design.* Minneapolis: Fortress Press, 2006.
> Written for the general public, this thin book (116 pages) gives reasoned objections to the intelligent design movement. One of the more interesting figures in theistic evolution, Ayala combines a traditional religious background with an international reputation as an evolutionary scientist.

Ayala, Francisco J. *Darwin's Gift to Science and Religion.* Washington, DC: Joseph Henry Press, 2007.
> Darwin's system appeared to eliminate design, purpose, and progress from science, and Ayala believes such issues should be left to theology, which is best equipped to deal with them. He also suggests that Darwin's contribution to religion may be equal to his contribution to science. Ayala believes that Darwinian

evolution provides the best answer to the perennial problem of theodicy, how evil can exist in a world created by an omniscient, benevolent God.

Allen, Garland. *Life Sciences in the Twentieth Century.* Cambridge: Cambridge University Press, 1978.

A good overview of the enormous scientific developments from 1900 to 1978, with the neo-Darwinian synthesis being perhaps the most important.

Appleman, Philip, ed. *Darwin.* New York: W.W. Norton, 1970.

A Norton casebook, with basic selections from Darwin's own writings, in addition to commentaries and essays on the influence of evolutionary theories on all aspects of culture. Especially recommended for student research.

Armstrong, Karen. *A History of God: The 4000-Year Quest of Judaism, Christianity, and Islam.* New York: Alfred A. Knopf, 1993.

A former Roman Catholic nun, Armstrong examines ways the Deity is viewed by philosophers, mystics, and reformers in the three Abrahamic religious traditions. The book's last two chapters are especially relevant, examining the "death of God" proclaimed by postmoderns and asking questions about the future of belief.

Ball, William Bentley. *Mere Creatures of the State?: A View from the Courtroom.* Notre Dame, IN: Crisis Books, 1994.

A distinguished lawyer surveys church-state legal interactions, spanning several decades. His thesis is that religion has slipped from its first place in U.S. freedoms, and he evaluates the legal decisions that have been responsible.

Bates, Stephen. *Battleground: One Mother's Crusade, the Religious Right, and the Struggle for Control of Our Classrooms.* New York: Henry Holt, 1993.

A discussion of events in Church Hill, Tennessee, and the personal story of Vicki Front's concern over the textbooks used by her child. The control of public school textbooks remains one of the central issues in the evolution-creation struggle.

Beckwith, Francis J. *Law, Darwinism, and Public Education.* Lanham, MD: Rowman & Littlefield, 2003.

In one of the most comprehensive treatments of the subject, Beckwith's sympathies are with intelligent design.

Behe, Michael. *Darwin's Black Box: The Biochemical Challenge to Evolution.* New York: Free Press, 1996.

One of the strongest defenders of intelligent design, Behe is an academic scientist who bases his viewpoint chiefly on his concept of irreducible complexity.

Behe, Michael. *The Edge of Evolution: The Search for the Limits of Darwinism.* New York: Free Press, 2007.

Based on anatomical, genetic, and fossil evidence, Behe accepts the principle that all species on earth descended from a common ancestor but further argues that the entire universe is fine-tuned for human life and must have been intelligently designed.

Bowler, Peter. *The Eclipse of Darwinism.* Baltimore: Johns Hopkins University Press, 1992.

An examination of problems in classic Darwinism and the emergence of neo-Darwinism in the early 20th century, with a good overview of evolutionary developments throughout the century.

Bowler, Peter. *Evolution: The History of an Idea*. Berkeley: University of California Press, 1984.

A detailed historical survey of the development of evolutionary theory. This book is frequently used in history of science college courses.

Bowler, Peter. *Monkey Trials and Gorilla Sermons: Evolution and Christianity from Darwin to Intelligent Design*. Cambridge, MA: Harvard University Press, 2007.

An examination of both religious interpretations of evolution and corrections in Darwinism since its beginnings.

Brooke, John Hedley. *Science and Religion: Some Historical Perspectives*. Cambridge: Cambridge University Press, 1991.

A survey of the interactions between science and Christianity since the 16th century. Of special interest are the sections on natural theology and evolution.

Brown, Janet. *Charles Darwin: The Power of Place*. New York: Knopf, 2002.

An exhaustive biography of Darwin from the publication of *On the Origin of Species* to his death, placing his life and thought within the context of his time.

Burtchaell, James T. *The Dying of the Light: The Disengagement of Colleges and Universities from Their Christian Churches*. Grand Rapids, MI: Eerdmans, 1998.

Although the emphasis here is not on public schools, the historical approach to higher education is insightful. Burtchaell explains how great universities founded by leading Protestant denominations have gradually abandoned their religious roots.

Cadden, John J., and Patrick R. Brostowin, eds. *Science and Literature, A Reader*. Boston: D. C. Heath, 1964.

A readable anthology of basic literary selections responding to the problem of "the two cultures" as delineated by C. P. Snow.

Campbell, John Angus, and Stephen C. Meyer. *Darwinism, Design, and Public Education*. East Lansing: Michigan State University Press, 2003.

A helpful collection of essays on intelligent design, American pluralism, and the public schools. The authors examine biology textbooks and ask the question: Do religious implications turn a theory into religion?

Campbell, John Angus, and Stephen C. Meyer, eds. *Darwinism, Design, and Public Education*. East Lansing: Michigan State University Press, 2003.

A balanced treatment, bringing together one of the best collections of essays on intelligent design in public education.

Cantor, Geoffrey, and Marc Switlitz, eds. *Jewish Tradition and the Challenge of Darwinism*. Chicago: University of Chicago Press, 2006.

A rare collection of fine essays on the Jewish engagement with evolutionary theories that challenge traditional beliefs. Of special interest are the discussions of the impact of Darwinism on Zionism and anti-Semitism. One essay evaluates the work of Gerald L. Schroeder and other modern Orthodox thinkers. Several viewpoints are expressed.

Carlisle, Christopher, with W. Thomas Smith, Jr. *Understanding Intelligent Design*. Indianapolis, IN: Alpha Books, 2006.

Although this is a volume in the Complete Idiot's Guide series, which might easily be overlooked, it provides a good introduction for a beginning student.

The author is an Episcopal priest with a fair, well-balanced presentation of the subject. Of special value are the glossary, the bibliography, and the timeline.

Cartwright, John H., and Brian Baker, eds. *Literature and Science: Social Impact and Interaction.* Santa Barbara, CA: ABC-CLIO, 2005.

A comprehensive anthology of major sources from medieval to modern times, with good historical commentary. One of the best books available, especially for students exploring the social impact of evolution.

Chappell, Dorothy F., and E. David Cook. *Not Just Science: Where Christian Faith and Natural Science Intersect.* Grand Rapids, MI: Zondervan, 2005.

The work of two scientists of Christian persuasion, this book presents the natural world as a way, even today, of understanding God and informing faith. An especially good discussion of Christian values and ethics as they relate to the applied sciences is included.

Colling, Richard G. *Random Designer.* Bourbonnais, IL: Browning Press, 2004.

A conservative Christian who is also a leading microbiologist, Collins criticizes the intelligent design movement and suggests ways his own religion and his science neither interfere nor conflict with one another.

Collins, Francis S. *The Language of God.* New York: Free Press, 2006.

An essential book for theistic evolution, written by the head of the Human Genome Project, explaining how his science helped lead him to his Christian faith.

Conkin, Paul K. *When All the Gods Trembled: Darwinism, Scopes, and American Intellectuals.* Lanham, MD: Rowman & Littlefield, 1998.

A highly readable examination of the impact of evolutionary thinking on U.S. religious thought by a Vanderbilt University professor whose strong opinions are always expressed with liveliness.

Copan, Paul, and William Lane Craig. *Creation out of Nothing: A Biblical, Philosophical, and Scientific Exploration.* Grand Rapids, MI: Baker Academic, 2004.

A philosopher and a theologian examine Old Testament, New Testament, extrabiblical, philosophical, and scientific evidences for the doctrine of creation ex nihilo. The last chapter explores naturalistic alternatives to this belief. Some technical background is helpful though not absolutely essential in following their arguments.

Coulter, Ann. *Godless: The Church of Liberalism.* New York: Crown Forum, 2006.

A conservative satirist, television entertainer, and social critic portrays liberalism as a godless religion that is tainting all of American society, especially education. Several sections deal with what she believes are the fallacies of Darwinism, especially the chapter titled "Proof for How the Walkman Evolved into the iPod by Random Mutation." Coulter has been a number-one author on the *New York Times* best-seller list, and her books are lively, entertaining, opinionated, and provocative.

Coyne, Jerry A. *Why Evolution Is True.* New York: Viking, 2009.

A strong case for evolution without intelligent design, expressed in a clear, readable style. This is the book that Richard Dawkins claimed would convert any reasonable person away from intelligent design and "its country cousin," young earth creationism. Of special interest is chapter eight, "What about U.S.?"

Curtis, Gregory. *The Cave Painters: Probing the Mysteries of the World's First Artists.* New York: Alfred A. Knopf, 2006.

    Written in a suspenseful narrative style, this examination of cave art in France is informative and throws new light on the creative potential of early humans.

Darwin, Charles. *On the Origin of Species.* New York: Athenaeum, 1967.

    Because it is always helpful to return to sources, this facsimile of the first edition of Darwin's classic work, from 1859, is basic.

Davidson, Keay. *Carl Sagan, A Life.* New York: John Wiley, 1999.

    A readable biography of the leading science popularizer of the last half of the 20th century. Of special interest to those who believe in the likelihood of extraterrestrial intelligence.

Davies, Paul. *The Fifth Miracle: The Search for the Origin and Meaning of Life.* New York: Simon & Schuster, 1999.

    A readable examination of the nature of life, its probable origin, and the possibility that it has always existed in the universe. Of special interest to those who speculate on potential extraterrestrial intelligence.

Davies, Paul. *God and the New Physics.* New York: Simon & Schuster. 1983.

    A clear introduction to relativity and quantum theory and their implications for theism; a provocative book requiring careful attention rather than advanced scientific education.

Davies, Paul. *The Mind of God: The Scientific Basis for a Rational World.* New York: Simon & Schuster, 1992.

    Despite the audacity of his title, Davies, who describes himself as a theist who follows no conventional religion, provides a valuable discussion of both the strengths and limits of science.

Dawkins, Richard. *The Blind Watchmaker: Why the Evidence of Evolution Reveals a Universe without Design.* New York: W.W. Norton, 1986.

    Highly readable and witty, as are all of Dawkins's books, this presentation of evolution is from an aggressively atheistic standpoint.

Dawkins, Richard. *The God Delusion.* Boston: Houghton Mifflin, 2006.

    An attack from a scientific and humanistic point of view on religion in general and Christianity in particular. According to Dawkins, it was Darwin who made atheism inevitable. Though he believes Christianity is "stupid and insane," he charitably acknowledges that it is not illegal!

Dawkins, Richard. *The Selfish Gene.* Oxford: Oxford University Press, 1976.

    In this popular presentation, Dawkins views evolution as operating primarily at the gene level. With his usual felicity of style, Dawkins presents his theory that the gene is the significant unit of natural selection in both humans and lower animals.

DelFattore, Joan. *The Fourth R: Conflicts over Religion in America's Public Schools.* New Haven, CT: Yale University Press, 2004.

    In a dramatic, witty style, DelFattore traces school-prayer battles and other religious observances from the early 1800s to the present, demonstrating how majority rule and individual rights attempt to work themselves out in these conflicts.

Dembski, William A. *The Design Revolution: Answering the Toughest Questions about Intelligent Design.* Downers Grove, IL: InterVarsity Press, 2004.

   A basic introduction to the intelligent design movement, by one of its leaders.

Dembski, William A. *Intelligent Design: The Bridge between Science and Theology.* owners Grove, IL: InterVarsity Press, 1999.

   In this highly detailed and somewhat technical book, Dembski presents his strongest scientific case for intelligent design.

Dembski, William A. *No Free Lunch: Why Specified Complexity Cannot Be Purchased without Intelligence.* Lanham, MD: Rowman & Littlefield, 2001.

   An approach to intelligent design from a mathematical and philosophical position. A technical book, but a basic source for the movement.

Dembski, William A., ed. *Uncommon Dissent: Intellectuals Who Find Darwinism Unconvincing.* Wilmington, DE: ISI Books, 2004.

   A compilation of anti-Darwinian materials from a variety of personalities important to the intelligent design movement.

Dembski, William A., Wayne J. Downs, and Fr. Justin B. A. Frederick, eds. *The Patristic Understanding of Creation: An Anthology of Writings from the Church Fathers on Creation and Design.* Riesel, TX: Erasmus Press, 2008.

   A pertinent collection of creation writings from both Greek and Latin fathers of the Christian Church, especially useful for a historical understanding of the evolution–creation controversy.

Dembski, William A., and Michael Ruse, eds. *Debating Design from Darwin to DNA.* Cambridge: Cambridge University Press, 2004.

   Though Dembski and Ruse hold different points of view, they have brought together a stimulating collection examining major issues in the intelligent design–neo-Darwinian controversy.

Dennett, Daniel C. *Breaking the Spell: Religion as a Natural Phenomenon.* New York: Viking, 2006.

   An investigation of religion from a skeptical, scientific point of view, Dennett explores possible psychological and cultural origins of religion. The book is wordy, repetitive, and excessively discursive but does contain provocative facts and ideas.

Denton, Michael. *Evolution: A Theory in Crisis.* Bethesda, MD: Adler & Adler, 1985.

   In a basic source for the intelligent design movement, Denton identifies gaps in evolutionary theory and unanswered questions.

Desmond, Adrian, and James Moore. *Darwin: The Life of a Tormented Evolutionist.* New York: Time Warner, 1991.

   An interesting account of Darwin's life and personal trials against the backdrop of his times, stressing the problems that Darwin faced as he long delayed presenting his theory because of social and family concerns.

Diamond, Jared. *The Third Chimpanzee: The Evolution and Future of the Human Animal.* New York: HarperCollins, 1992.

   A diverting look at human culture from the viewpoint of evolution, exploring how the small genetic differences between humans and chimpanzees have made possible the creation of civilizations, religions, multiple languages, arts, and science.

Dobzhansky, Theodosius. *Mankind Evolving.* New York: Bantam Books, 1969.

One of the 20th century's most distinguished geneticists explains how human evolution must be understood through the interaction of the biological and cultural. The last chapters express Dobzhansky's admiration of the mysticism of Teilhard de Chardin.

Durham, James R. *Secular Darkness: Religious Right Involvement in Texas Public Education, 1963–1989.* New York: Lang, 1995.

Durham reviews the battles in the Texas public schools over religion and science. His sympathies lie with the scientific establishment.

Eldredge, Niles. *Darwin: Discovering the Tree of Life.* New York: W.W. Norton, 2005.

In a beautifully illustrated book by one of today's leading evolutionary thinkers, special attention is given to Darwin's notebooks and the development of his germinal ideas. The last chapter looks at creationism in the 21st century. Eldredge, along with Stephen Jay Gould, is one of the developers of the punctuated equilibria theory of evolution.

Ferguson, John. *Clement of Alexandria.* New York: Twayne, 1974.

A useful introduction to this important Greek Church father, with some attention to the ways Clement and other early theologians interpreted the Bible. The summary of the ways Greek pagan philosophy influenced early Christian theology is especially revealing. The book follows a format that makes basic information retrieval easy.

Ferngren, Gary B., ed. *The History of Science and Religion in the Western Tradition: An Encyclopedia.* New York: Garland, 2000.

A valuable resource of 100-plus articles, this reference work examines many sides of the issue. Chapters feature such topics as medieval science and religion, Islamic influences, the Copernican revolution, the problems of Galileo, early Protestant attitudes toward science, Newtonian physics, natural theology, geology and paleontology, Darwinism, cosmogonies, as well as Roman Catholic, evangelical, and fundamentalist attitudes toward science. Edward J. Lawson's chapter on the Scopes trial is basic, as is the section on intelligent design. The chapters on gender, social construction, and postmodern approaches to science introduce new issues waiting to be explored.

Foerst, Anne. *God in the Machine: What Robots Teach Us about Humanity and God.* New York: Dutton, 2004.

Despite its intriguing title, the book delivers few new ideas. The author, who works with humanoid robots, has been called the country's only "robotics theologian." This topic might well inspire explorations by highly creative students.

Frankenberry, Nancy K., ed. *The Faith of Scientists, In Their Own Words.* Princeton, NJ: Princeton University Press, 2008.

Excerpts from the relevant writings of scientists from Galileo through Ursula Goodenough, including both conventional believers and atheists. Commentary by the editor is illuminating.

Fraser, James W. *Between Church and State: Religion and Public Education in a Multicultural America.* New York: St. Martin's Press, 1999.

After examining recent approaches to religion in education, Fraser argues that religion should be a part of a multicultural curriculum. He provides numerous examples and gives renewed attention to global concerns.

Gaddy, Barbara B., et al. *School Wars: Resolving Our Conflicts over Religion and Values.* San Francisco: Jossey-Bass, 1996.

The essays in this collection cover most of the religious issues in public schools, though the problems are more vivid than the suggested solutions are convincing.

Geisler, Norman. *Creation and the Courts: Eighty Years of Conflict in the Classroom and the Courtroom.* Wheaton, IL: Crossway Books, 2007.

This examination of major court cases, beginning with the *State v. John Scopes* trial in 1925, is a basic source. Geisler, a conservative Bible scholar and distinguished professor at Southern Evangelical Seminary, was an expert witness for the state in the *McLean v. Arkansas Board of Education* trial in Arkansas and is a knowledgeable defender of theistic approaches to human origins.

Gibbons, Ann. *The First Human: The Race to Discover Our Earliest Ancestors.* New York: Doubleday, 2006.

A dynamic narrative of recent developments in paleontology describing the trials of fossil hunters in Africa who struggle to find missing links between modern humans and their early ancestors.

Gilbert, James. *Redeeming Culture: American Religion in an Age of Science.* Chicago: University of Chicago Press, 1997.

An intellectual-cultural history that examines the contentions between modern science and religion, beginning with the Scopes trial. The chapter on William Jennings Bryan is especially pertinent.

Gilkey, Langdon. *Creationism on Trial: Evolution and God at Little Rock.* San Francisco: Harper & Row, 1985.

Gilkey, a well-regarded theologian and professor of theology at the University of Chicago Divinity School, gives a readable firsthand account of the 1981 *McLean v. Arkansas Board of Education* trial in Little Rock. A witness for the prosecution, he feared the intelligent design movement, which he believed to be a tool of a fundamentalist religion with plans to control U.S. politics and society.

Gillispie, Charles Coulston. *Genesis and Geology: The Impact of Scientific Discoveries upon Religious Beliefs in the Decades before Darwin.* New York: Harper & Row, 1951.

A survey of developments in geology that led up to Darwin's discoveries, examining scientific advances and religious responses to them.

Ginger, Ray. *Six Days or Forever?* New York: Signet Books, 1958.

A narrative account of the *State v. John Scopes* trial in Dayton, Tennessee, notable for the biographical information it provides on William Jennings Bryan and Clarence Darrow. More depth is provided than in most other accounts of these events.

Gingerich, Owen. *God's Universe.* Cambridge, MA: Belknap Press of Harvard University Press, 2006.

This slight but valuable book is by an astronomer who is also a practicing Mennonite. Gingerich argues that the universe has intention and purpose, though he opposes the intelligent design movement, which he feels confuses science and religion.

Gleason, Philip. *Contending with Modernity: Catholic Higher Education in the Twentieth Century.* New York: Oxford University Press, 1995.

Though his concern is not with the public schools, Gleason identifies general problems of religious education in a secular society. He explains why many Roman Catholics have long been skeptical of the U.S. public schools, which they believe have been dominated by Protestant values and influences.

Gonzalez, Guillermo, and Jay W. Richards. *The Privileged Planet: How Our Place in the Cosmos Is Designed for Discovery.* Washington, DC: Regnery, 2004.

A distinguished astronomer explores earth's favorable location in the cosmos, making space discovery and exploration possible and likely in the near future.

Gould, Stephen Jay. *Rocks of Ages: Science and Religion in the Fullness of Life.* New York: Ballantine, 1999.

Gould's strongest statement of his doctrine of the two magisteria, science and religion, each with its separate province. Although Gould makes a generous case, he does not fully acknowledge that both magisteria have frequently overlapped in the past and continue to challenge each other in the present.

Gould, Stephen Jay. *The Structure of Evolutionary Theory.* Cambridge, MA Belknap Press of Harvard University Press, 2002.

The final work of a major evolutionary scientist, an in-depth treatment of evolution, reviewing newer discoveries and theories that enhance and modify Darwinism.

Gould, Stephen Jay. *Wonderful Life: The Burgess Shale and the Nature of History.* New York: W.W. Norton, 1989.

A clear presentation of Gould's theory of punctuated equilibria for the nonspecialist. Highly readable and informative.

Greene, John C. *The Death of Adam: Evolution and Its Impact on Western Thought.* Ames: Iowa State University Press, 1959.

A classic account of the background of Darwin's theories and their later impact.

Gregory, Frederick. *Natural Science in Western History.* Boston: Houghton Mifflin, 2008.

A balanced survey of scientific inquiry from antiquity to the present, by one of the leading science historians of our time.

Gross, Paul, and Norman Levitt, eds. *The Higher Superstition.* Baltimore: Johns Hopkins University Press, 1997.

A critical examination of postmodern thought in its application to science. The editors suggest that postmodernism may be as strong a challenge to Darwinism as intelligent design.

Gunn, Angus M. *Intelligent Design and Fundamentalist Opposition to Evolution.* Jefferson, NC: McFarland, 2006.

An analysis of creationism and intelligent design as they come into contact with evolutionary theory and modern science. Gunn reviews the history of fundamentalism, in reality a modern movement, as it conflicts with both evolution and traditional biblical scholarship.

Hall, Kermit L., editor-in-chief. *The Oxford Companion to the Supreme Court of the United States.* New York: Oxford University Press, 1992.

This comprehensive reference book on the history and deliberations of the Supreme Court contains helpful sections on religion and the law and education and the law. There is also a concise review of relevant court cases and decisions.

Hamer, Dean. *The God Gene: How Faith Is Hardwired into Our Genes.* New York: Doubleday, 2004.

A much-discussed and debated book that presents a leading geneticist's thesis that religious feelings are part of the human genetic makeup. Hamer also explores the ways these feelings may assist in human survival.

Haught, John. *God after Darwin.* Boulder, CO: Westview Press, 2000.

A Roman Catholic approach to evolutionary science, from a Georgetown University theology professor. Less reliant on the Bible than are conservative Protestants, Haught shows how Roman Catholics have generally been much more open to evolutionary ideas.

Haynes, Charles C., and Oliver Thomas. *Finding Common Ground: A First Amendment Guide to Religion and Public Schools.* Nashville, TN: First Amendment Center, 2007.

A useful manual offering historical perspective, instructional guides, and full discussions of issues relating to religious liberty and the public schools.

Hefner, Philip. *The Human Factor: Evolution, Culture, and Religion.* Minneapolis: Fortress Press, 1993.

A theological accommodation of Christianity to Darwinian theory.

Hitchens, Christopher. *Darwin and the Science of Evolution.* New York: Twelve Hachette Book Group, 2007.

Hitchens is not a scientist but considers himself a spokesperson for "a new Enlightenment." An avowed atheist, he angrily denounces religion and religious heroes such as Muhammad. A frequent critic of such public personalities as Princess Diana and Mother Teresa, Hitchens has rarely encountered an individual he likes. He sees no redeeming features in religion and feels that religious indoctrination of the young is a form of child abuse. He attacks Judaism, Christianity, and Islam equally, though he retains a special animosity for Christianity, the religion most familiar to him.

Hoeveler, J. David. *The Evolutionists: American Thinkers Confront Charles Darwin, 1860–1920.* New York: Rowman & Littlefield, 2007.

In his review of 14 theologians and scientists as they have confronted Darwin's views on human evolution, Hoeveler scrutinizes the influence of Darwin on both social and religious thought.

Hofstadter, Richard. *Social Darwinism in American Thought*. Boston: Beacon Press, 1992.
  A later edition of a book that first appeared in 1955, providing the most thorough examination of social Darwinism in the United States.
Horvitz, Leslie Alan. *Evolution*. Indianapolis, IN: Alpha Books, 2002.
  A volume in the Complete Idiot's Guide series, *Evolution* offers a lively, informed presentation of a complicated subject, by an experienced science writer. The glossary and timeline are useful to students, particularly college undergraduates.
Howell, Kenneth J. *God's Two Books: Copernican Cosmology and Biblical Interpretation in Early Modern Science*. Notre Dame, IN: University of Notre Dame Press, 2007.
  A survey of Catholic and Protestant biblical interpretations of the 16th and 17th centuries and their relationship to the science of their times.
Humes, Edward. *Monkey Girl: Evolution, Education, Religion, and the Battle for America's Soul*. New York: HarperCollins, 2007.
  A readable book with a lot of information but an unfortunate tendency to demonize intelligent design proponents.
Hunter, Cornelius G. *Darwin's God: Evolution and the Problem of Evil*. Grand Rapids, MI: Brazos Press, 2001.
  From an intelligent design perspective, an examination of Darwinism and what light it does or does not shed on the problem of suffering and evil.
Israel, Charles A. *Before Scopes: Evangelicalism, Education, and Evolution in Tennessee, 1870–1925*. Athens: University of Georgia Press, 2004.
  A noncondescending discussion of the attitudes in Tennessee that led to the Scopes trial. Israel, an assistant professor of history at the University of the South in Sewanee, Tennessee, provides essential social context.
Jaravsky, David. *The Lysenko Affair*. Cambridge, MA: Harvard University Press, 1970.
  An engrossing account of how scientific dogmatism took root in the Soviet Union with devastating results. An object lesson in government interference with science.
Johnson, Phillip E. *Darwin on Trial*. Downers Grove, IL: InterVarsity Press, 1991.
  Possibly the germinal book of the intelligent design movement and a record of the conclusions of one of America's leading legal scholars. After careful scrutiny of Darwinism, Johnson found the evidence too slim to hold up in a U.S. court. Unlike most of the leading champions of intelligent design, Johnson is a law professor rather than a scientist.
Jordan, Paul. *Neanderthal: Neanderthal Man and the Story of Human Origins*. Phoenix Mill, UK: Sutton, 2000.
  An attractively illustrated introduction to Neanderthals, with informed speculations about their relationship to modern humans.
Jurinski, James John. *Religion in the Schools: A Reference Handbook*. Santa Barbara, CA: ABC-CLIO, 1998.
  Jurinski provides a chronology of his subject up to 1998, in addition to biographical sketches of important personalities. He also covers major documents,

excerpts from court cases, and critical examinations of basic issues. An invaluable source for the student.

Kevles, Daniel. *In the Name of Eugenics.* Cambridge, MA: Harvard University Press, 1998.

Kevles makes clear the social and political issues that led to the eugenics movement at the dawn of the 20th century. He also discusses its later ramifications.

King, Barbara J. *Evolving God.* New York: Doubleday, 2007.

Although lacking in clear focus and highly speculative, King, an evolutionary anthropologist, introduces a tantalizing subject that demands more attention. King is a popular lecturer and professor at the College of William and Mary.

King, Barbara J., ed. *The Origins of Language: What Nonhuman Primates Can Tell Us.* Santa Fe, NM: School of American Research Press, 1999.

A collection of essays by evolutionary anthropologists that speculate on human language by observing patterns of communication among other primates.

King, Thomas M., and Mary Wood Gilbert, eds. *The Letters of Teilhard de Chardin and Lucile Swan.* Washington, DC: Georgetown University Press, 1993.

In letters of the Jesuit paleontologist to his platonic friend of many years, Teilhard outlines many of his central ideas.

King, Ursula. *Towards a New Mysticism: Teilhard de Chardin and Eastern Religions.* New York: Seabury Press, 1981.

King analyzes the spirituality of the famous French Jesuit paleontologist who sought to reconcile Christianity and evolution. She evaluates the contention of many readers of Teilhard's books that his mysticism is more akin to Indian religions than to Christianity.

Kitcher, Philip. *Abusing Science, The Case against Creationism.* Cambridge, MA: The MIT Press, 1982.

A lively refutation of creationism, identified as both a social and political movement, and a "manual for intellectual self-defense." Kitcher examines the literature and strategy of the creationist movement and advises ways of refuting its claims. A readable polemic.

Koestler, Arthur. *The Case of the Midwife Toad.* New York: Random House, 1972.

An intriguing portrait of Paul Kammerer, one of the most romantic, interesting, and tragic figures in evolutionary science. Exceptionally well written and readable.

Kragh, Helge. *Cosmology and Controversy: The Historical Development of Two Theories of the Universe.* Princeton, NJ: Princeton University Press, 1996.

A comparison of the steady-state versus Big Bang theories of the origin of the universe. Although the chief interest is scientific, some theological implications are suggested.

Küng, Hans. *The Beginning of All Things: Science and Religion.* Translated by John Bowden. Grand Rapids, MI: Eerdmans, 2007.

A controversial European Roman Catholic (who is no longer allowed to call himself an official Catholic theologian) takes issue with both dogmatic scientists and closed-minded religionists. Küng believes that enlightened religion can be reconciled with valid science.

Larson, Edward J. *Evolution: The Remarkable History of a Scientific Theory.* New York: Random House, 2004.

Like all Larson's books, this one is both entertaining and informative, with an especially thorough treatment of America's anti-Darwin crusade.

Larson, Edward J. *Evolution's Workshop: God and Science on the Galapagos Islands.* New York: Basic Books, 2001.

Larson reviews the fieldwork that has taken place in the Galapagos Islands, beginning with Darwin and continuing with contemporary evolutionary biologists.

Larson, Edward J. *Sex, Race and Science: Eugenics in the Deep South.* Baltimore: Johns Hopkins University Press, 1995.

While providing a good introduction to the subject of eugenics, Larson's concentration is on its sad story in the American South.

Larson, Edward J. *Summer for the Gods: The Scopes Trial and America's Continuing Debate over Science and Religion.* Cambridge, MA: Harvard University Press, 1997.

The best book on the Scopes trial, in which Larson corrects many common misconceptions. He combines his fields of history and law with a keen understanding of the importance of Darwinian biology.

Larson, Edward J. *Trial and Error: The American Controversy over Creation and Evolution.* New York: Oxford University Press, 1985.

Larson examines legal maneuvers from 1920 through 1982. Possibly the most readable book on this subject.

Leeming, David Adams, with Margaret Adams Leeming, eds. *A Dictionary of Creation Myths.* New York: Oxford University Press, 1994.

In this attractive reference book, the Leemings bring together creation stories from all over the world and from all periods of history. Of special interest are the accounts from the living world religions.

Levine, George, and Owen Thomas, eds. *The Scientist vs. the Humanist.* New York: W.W. Norton, 1963.

An anthology of essays, poetry, and even chapters from fictional works examining what C. P. Snow called "the two cultures."

Lindberg, David C., and Ronald L. Numbers, eds. *God and Nature: Historical Essays on the Encounter between Christianity and Science.* Berkeley: University of California Press, 1986.

A selection of essays from a variety of writers, from ancient times to the present. Five chapters deal with evolution and religion.

Livingstone, David. *Darwin's Forgotten Defenders.* Grand Rapids, MI: Eerdmans, 1987.

An identification and evaluation of conservative theologians, both before and after Darwin, who did not find evolutionary ideas offensive.

Lubac, Henri de, S. J. *Teilhard de Chardin: The Man and His Meaning.* Translated by René Hague. New York: Hawthorne Books, 1965.

A full introduction to Teilhard's religious and scientific thought by a fellow French Jesuit, who was the world's leading Teilhard scholar.

Marsden, George M. *Fundamentalism and American Culture.* Oxford: Oxford University Press, 1980.

Considered the standard study of religious fundamentalism, Marsden charts its rise in the latter part of the 19th century to its flourishing in the 20th century.

Marty, Martin E., with Jonathan Moore. *Education, Religion, and the Common Good: Advancing a Distinctly American Conversation about Religion's Role in Our Shared Life.* San Francisco: Jossey-Bass, 2000.

Marty, a genial professor emeritus of the University of Chicago and contributing editor of *Christian Century,* has an extraordinary talent for stating the obvious at great length. The chief value of this book is its fine list of suggested readings dealing with religion in the public schools.

McGrath, Alister, and Joanna Collicutt McGrath. *The Dawkins Delusion/Atheist Fundamentalism and the Denial of the Divine.* Downers Grove, IL: InterVarsity Press, 2007.

A slight but intelligent book designating Richard Dawkins an "atheistic fundamentalist" and pointing out fallacies in his thinking. The authors are British, like their subject, and have specialized in psychology and molecular biophysical development as well as Christian thought.

Miller, Kenneth R. *Finding Darwin's God: A Scientist's Search for Common Ground between God and Evolution.* New York: Harper Perennial, 2000.

A thoughtful examination of evolution and religion by a Roman Catholic evolutionist. The title is a bit tricky, but Miller explains his meaning cogently.

Miller, Kenneth R. *Only a Theory: Evolution and the Battle for America's Soul.* New York: Viking, 2008.

This professor of biology at Brown University and Roman Catholic scholar takes issue with the intelligent design movement and makes the argument for stronger science education in the United States.

Montagu, Ashley, ed. *Science and Creationism.* New York: Oxford University Press, 1984.

A collection of basic essays by Isaac Asimov, George M. Marsden, Stephen Jay Gould, Kenneth R. Miller, and others, critical of creationism.

Moore, James R. *The Post-Darwinian Controversies; A Study of the Protestant Struggle to Come to Terms with Darwin in Great Britain and America 1870–1900.* London: Cambridge University Press, 1979.

An examination of Darwinism in transition, the challenge of Lamarckian genetics, Christian Darwinism, and Darwin's relevance to orthodox religious theology.

Moran, Jeffrey P. *The Scopes Trial: A Brief History with Documents.* New York: Palgrave Macmillan, 2002.

An essential resource containing transcripts from the trial along with documents detailing the publicity surrounding it. The introduction and commentaries are especially helpful in understanding the social and religious forces behind the Scopes trial, still pertinent as a foundation for an informed awareness of the intelligent design controversies.

Morris, Henry M. *History of Modern Creationism.* Santee, CA: Institute of Creation Research, updated edition, 1993.

A history of the creation science movement by one of its founders.

Nash, Robert J. *Faith, Hype, and Clarity: Teaching about Religion in American Schools and Colleges.* New York: Teachers College Press, 1999.

Nash suggests that religious instruction be considered in four categories: fundamentalist, prophetic, alternate spiritualities, and post-theistic.

Nettancourt, Dreux de. *Tant ou'il y aura des atomes.* Brussels: Le Cri édition, 2002.

This very useful book by a leading European scientist is available only in French. The clear, concise presentation of the latest scientific findings on human origins is without equal for the student. Especially interesting is Nettancourt's highly personal chapter on religion, though his admiration of Father Teilhard de Chardin appears uncritical and excessive.

Noddings, Ned. *Education for Intelligent Belief or Unbelief.* New York: Teachers College Press, 1993.

Noddings believes that religious topics should be included in the public school curriculum and stresses the necessity of impartial teachers who will encourage critical thinking.

Nord, Warren A. *Religion and American Education: Rethinking a National Dilemma.* Chapel Hill: University of North Carolina Press, 1995.

A good overview of the subject with an especially helpful section on the importance of religious knowledge to liberal education.

Nord, Warren A., and Charles C. Haynes. *Taking Religion Seriously across the Curriculum.* Nashville, TN: First Amendment Center, 1998.

While agreeing that public schools must not proselytize, the authors make a strong case that religious studies must not be ignored in kindergarten through 12th grade. They offer suggestions of ways religion may be integrated into the other subjects that form the curriculum.

Numbers, Ronald L. *Darwinism Comes to America.* Cambridge, MA: Harvard University Press, 1998.

A basic college course book, which chronicles scientific and religious responses to Darwinism in the United States.

Numbers, Ronald L. *The Creationists: The Evolution of Scientific Creationism.* New York: Alfred A. Knopf, 1992.

The definitive history of the creation science movement, with the career of George McCready Price and his new "catastrophism" especially well covered.

O'Leary, Denyse. *By Design or Chance? The Growing Controversy on the Origins of Life in the Universe.* Minneapolis: Augsburg Fortress, 2004.

A provocative discussion, favorable to the intelligent design movement.

Orel, Vitèzslav. *Gregor Mendel: The First Geneticist.* Oxford: Oxford University Press, 1996.

A sound biography of Mendel, one of the most interesting figures in Western history and the discoverer of basic laws of inheritance.

Pearcey, Nancy. *Total Truth: Liberating Christianity from Its Cultural Captivity.* Wheaton, IL: Crossway Books, 2004.

A lively discussion of Christianity and its relationship to Darwinian science, promoting an "enlightened" Christianity that is not hostile to scientific research.

Peters, Ted, and Martinez Hewlett. *Evolution: From Creation to New Creation.* Nashville, TN: Abington Press, 2003.

A good summary of reactions to Darwin and the arguments of creationists, intelligent design theorists, and theistic evolutionists.

Peters, Ted, and Martinez Hewlett. *Theological and Scientific Commentary on Darwin's Origin of Species.* Nashville, TN: Abington Press, 2008.

The work of a theology professor and a molecular biologist, surveying the theological, philosophical, and social interpretations of Darwin during the past 150 years.

Phy, Allene Stuart, ed. *The Bible and Popular Culture in America.* Philadelphia and Chico, CA: Fortress Press and Scholars Press, 1985.

A germinal collection of essays relating the Bible to popular novels, visual art, the electronic church, children's literature, tall tales, humor, and traveling Bible salesmen. Of special interest is the essay by Charles Wolfe on country music, which discusses the genre's celebration of the Scopes trial.

Polkinghorne, John. *Belief in God in an Age of Science.* New Haven, CT: Yale University Press, 1998.

A distinguished physicist and Church of England theologian gives strong reasons for religious belief in the age of science.

Power, Edward J. *Religion and the Public Schools in 19th Century America: The Contribution of Orestes A. Brownson.* Mahwah, NJ: Paulist Press, 1996.

A good historical perspective that examines educational theories that emphasize the compatibility of democracy with Roman Catholicism.

Provenzo, Eugene F., Jr. *Religious Fundamentalism and American Education: The Battle for the Public Schools.* Albany: State University of New York Press, 1990.

Provenzo examines the conflict between secularization in the public schools and the desires of those he feels are ultrareligious parents. He examines the proposition, often voiced, that a sect of Christianity is attempting to control the public schools.

Radosh, Daniel. *Rapture Ready!* New York: Scribner, 2008.

Radosh leads a witty tour through the world of religious fundamentalism. Although filled with detail, this book should be read primarily for pleasure. Of special interest is the description of the creation museums he visits.

Rana, Fazale. *The Cell's Design: How Chemistry Reveals the Creator's Artistry.* Grand Rapids, MI: Baker Books, 2008.

An intelligent design advocate who holds a PhD in chemistry, Rana explores the complexity of the cell, leading him to argue for supernatural design.

Ratzsch, Del. *Nature, Design, and Science: The Status of Design in Natural Science.* New York: State University of New York Press, 2001.

A well-informed introduction to the intelligent design movement, what it precisely means, and what it wishes to accomplish.

Regal, Brian. *Human Evolution: A Guide to the Debates.* Santa Barbara, CA: ABC CLIO, 2004.

A detailed account of the major scientific controversies surrounding Darwinism, this volume is additionally valuable for the inclusion of many resources for further study.

Roll-Hansen, Nils. *The Lysenko Effect: The Politics of Science.* Amherst, NY: Humanity Books, 2005.

In recounting the Soviet misfortunes resulting from the Lysenko effect, Roll-Hansen warns of the dangers of a governmentally enforced scientific orthodoxy.

Rolston, Homes, III. *Science and Religion: A Critical Survey.* New York: Random House, 1987.

Despite its dense style, this book is valuable for its comprehensive interdisciplinary overview.

Ross, Hugh. *A Matter of Days.* Colorado Springs, CO: Navpress, 2004.

Ross, an evangelical Christian, rejects young earth and six-day creationism, as well as evolution. A different point of view.

Rudwick, Martin. *The Meaning of Fossils.* Chicago: University of Chicago Press, 1985.

A comprehensive history of the origin and development of the science of paleontology.

Ruse, Michael. *Darwin and Design: Does Evolution Have a Purpose?* Cambridge, MA: Harvard University Press, 2003.

An exploration of evolution by a philosopher and historian of science.

Ruse, Michael. *The Evolution-Creation Struggle.* Cambridge, MA: Harvard University Press, 2005.

In this contribution to the history of ideas, Ruse views evolution and creationism as rival religious responses to a crisis of faith.

Russel, Robert J. *Cosmology: From Alpha to Omega.* Minneapolis: Fortress Press, 2008.

Russel relates Judeo-Christian theology to quantum physics, asserting that Darwinian science does not undermine beliefs in Divine Providence.

Sagan, Carl. *The Varieties of Scientific Experience: A Personal View of the Search for God.* Edited by Ann Druyan. New York: Penguin Books, 2006.

In these wide-ranging essays, Sagan ruminates on intelligent extraterrestrial life, the nature of the sacred, creationism, and intelligent design.

Schroeder, Gerald L. *Genesis and the Big Bang: The Discovery of Harmony between Modern Science and the Bible.* New York: Bantam Books, 1990.

A provocative blending of the Bible, Talmud, and Jewish mysticism with modern science by an MIT-trained physicist now teaching, writing, and lecturing in Jerusalem. A modern Orthodox Jewish point of view.

Schroeder, Gerald L. *The Hidden Face of God: Science Reveals the Ultimate Truth.* New York: Simon & Schuster, 2001.

From the Weizmann Institute in Jerusalem, Schroeder continues to relate scientific theories and discoveries to the Bible, Talmudic tradition, and the Cabala. The basic message of the Bible, which Schroeder accepts, is that God cares about the universe and is directly involved in it.

Schroeder, Gerald L. *The Science of God: The Convergence of Scientific and Biblical Wisdom.* New York: Broadway Books, 1997.

This distinguished Jewish scientist continues to reconcile the findings of modern biochemists, paleontologists, astrophysicists, and quantum physicists with the Bible, the Talmud, and the Cabala, moving beyond his earlier book in both detail and conception.

Scopes, John T., and James Presley. *Center of the Story: Memoirs of John T. Scopes.* New York: Holt, Rinehart and Winston, 1967.

Near the end of his life, Scopes looks back on the youthful experience that made his name a household word, correcting much of the misinformation responsible for the myth of the "Monkey Trial." His impressions, charitable in every instance, of both William Jennings Bryan and Clarence Darrow are vivid. He also discusses the impact of the play and film *Inherit the Wind* and the liberties taken with history. Scopes explains his reasons for volunteering to be the test case in the trial and what he hopes has been accomplished.

Scott, Eugenie C. *Creationism vs. Evolution: An Introduction.* Westport, CT: Greenwood Press, 2004.

Written for schoolteachers facing the evolution-creation controversy, this survey helpfully clarifies major issues.

Scott, Eugenie C., and Glenn Branch. *Not in Our Classrooms: Why Intelligent Design Is Wrong for Our Schools.* Boston: Beacon Press, 2006.

The authors strongly attack intelligent design, with the argument that conservative religion seeks to undermine the scientific integrity of American public education system.

Sears, James, and James C. Carper, eds. *Curriculum, Religion, and Public Education: Conversations for an Enlarging Public Square.* New York: Teachers College Press, 1998.

The writers in this volume examine the debates over textbooks, values, sex education, religion, science, and outcome-based education in the public schools. Although the focus is on educational theory, the chapters provide a good picture of contemporary school controversies.

Sermonti, Giuseppe. *Why Is a Fly Not a Horse?* Seattle, WA: Discovery Institute Press, 2005.

One of Discovery Institute's well-circulated publications, explaining why the movement is so important.

Shanavas, T. O. *Creation and/or Evolution: An Islamic Perspective.* Philadelphia: Xlibris, 2005.

An Islamic reconciliation of religious tradition with modern science by a noted physician and student of Islamic history. Shanavas argues that Muslim thinkers during the Golden Age had already intuited many of the insights of modern science.

Shapiro, Robert. *Origins: A Skeptic's Guide to the Creation of Life on Earth.* New York: Summit Books, 1985.

A survey of ideas about the origin of life, with the last chapters focusing specifically on religion-science issues.

Shermer, Michael. *Why Darwin Matters: The Case against Intelligent Design.* New York: Henry Holt, 2006.

In this defense of Darwinism, Shermer argues that religion and science cannot validly contradict one another and that Christians and conservatives must accept evolution, despite several unsolved issues. He feels that intelligent design is a remarkably uncreative theory. Although he is preachy and sometimes simplistic, his survey of ethics and their relationship to evolutionary theory is well worth reading.

Stace, W. T. *Man against Darkness and Other Essays.* Pittsburgh, PA: University of Pittsburgh Press, 1967.

Fifteen essays for a general reader by a leading philosopher, remarkable for his stylistic grace, who boldly acknowledges the devastation the modern scientific outlook has caused religion/.

Teilhard de Chardin, Pierre. *Christianity and Evolution.* Translated by René Hague. New York: Harcourt Brace, 1969.

A collection of 19 pertinent essays by the French Jesuit, who was a leading paleontologist.

Teilhard de Chardin, Pierre. *The Phenomenon of Man.* Translated by Bernard Wall. New York: Harper, 1959.

The most important work of the Jesuit priest and paleontologist, *Phenomenon* is regarded by many as the most significant document of theistic evolution in the 20th century.

Thaxton, Charles, Walter L. Bradley, and Roger L. Olsen. *The Mystery of Life's Origin.* Dallas, TX: Lewis and Stanley, 1984.

An articulate argument in favor of intelligent design.

Tort, Patrick. *Darwin and the Science of Evolution.* New York: Harry N. Abrams, 2001.

A well-illustrated biography of Darwin with an overview of his theories, especially useful for beginners. Both Darwin's defenders and opponents receive attention.

Tourney, Christopher P. *God's Own Scientists: Creationists in a Secular World.* New Brunswick, NJ: Rutgers University Press, 1994.

A sociologist analyzes the defenders of scientific creationism and their revolt against the materialism of the modern world.

Towne, Margaret Gray. *Honest to Genesis: A Biblical and Scientific Challenge to Creationism.* Baltimore: Publish America, 2003.

Directed chiefly at Christian fundamentalists, Towne feels that both enlightened religion and science undermine creationism.

Weikart, Richard. *From Darwin to Hitler.* New York: Palgrave Macmillan, 2004.

An examination of the abuses of Darwinian thought by the Nazi propaganda machine, leading to brutal racial policies and the Holocaust.

Wells, Jonathan. *Icons of Evolution: Why Much of What We Teach about Evolution Is Wrong.* Washington, DC: Regnery, 2002.

An impassioned defense of intelligent design by a biologist and theologian.

Wells, Jonathan. *The Politically Incorrect Guide to Darwinism and Intelligent Design.* Washington, DC: Regnery, 2006.

In an easy-to-read format, this senior fellow from the Discovery Institute identifies scientific charts based on guesswork rather than evidence and other inaccuracies taught in public schools. He also outlines what he believes is scientific evidence for intelligent design and identifies scientists who are skeptical of Darwinism.

Whitcomb, John C., and Henry M. Morris. *The Genesis Flood.* Nutley, NJ: Presbyterian and Reformed Publishing, 1961.

A classic statement for those who accept the literal account of Creation in Genesis, often credited with reviving the antievolution movement.

Whitehead, Alfred North. *Science and the Modern World: Lowell Lectures, 1925.* New York: New American Library, 1948.

One of the 20th century's leading philosophers, Whitehead writes with grace and knowledge of the scientific developments since the 17th century. He thoughtfully considers their religious implications.

Wilker, Benjamin, and Jonathan Witt. *A Meaningful World: How the Arts and Sciences Reveal the Genius of Nature.* Downers Grove, IL: InterVarsity press, 2006.

An enthusiastic presentation of the intelligent design viewpoint, bringing together insights from the sciences and the arts.

Wilson, A. N. *God's Funeral.* New York: W.W. Norton, 1999.

A British scholar examines the rapid decline of religious faith by writers, artists, and intellectuals at the end of the 19th century. The spread of Darwinism is identified as only one factor in this wide cultural change.

Winnick, E. *A Jealous God: Science's Crusade against Religion.* Nashville, TN: Nelson Current, 2005.

With a strong warning of the dangers of "celebrity science," Winnick discusses troubling applications of science: neofertilization techniques, designer babies, and the "federalization of science." Distortions of history in the play and movie versions of *Inherit the Wind* are also identified.

Woodward, Thomas. *Doubts about Darwin: A History of Intelligent Design.* Grand Rapids, MI: Baker Books, 2004.

A discussion of the history and development of the intelligent design movement, identifying the reasons for its continuing popularity despite legal setbacks.

The World's Most Famous Court Trial: Tennessee Evolution Case. Cincinnati: National Book Company, 1971.

Available in most large libraries, this book presents a complete stenographic report of the *State v. John Scopes* trial, including speeches and arguments of attorneys that were not heard by the jury that convicted John Scopes. The analyses of primary sources are especially welcome.

Zimmer, Carl. *Smithsonian Intimate Guide to Human Origins.* New York: HarperCollins, 2005.

Although the style is dry, the fine illustrations make this a useful book for students.

OTHER MEDIA

Biological Anthropology: An Evolutionary Perspective. Chantilly, VA: The Teaching Company, 2002.

Award-winning professor Barbara J. King, of the College of William and Mary, offers this series of video lectures relating human conduct to the behavior of hominids and the great apes.

The Darwinian Revolution. Chantilly, VA: The Teaching Company, 2008.

In a series of video lectures, master teacher Frederick Gregory of the University of Florida presents the history of evolutionary theory and its social impact.

Expelled: No Intelligence Allowed. 2009. 95 min. U.S., Nathan Frankowski.

Ben Stein is featured in this film, which documents the enforcement of scientific orthodoxy in universities around the world. Stein discovers that educators who do not conform to current politically correct scientific views are slandered, refused tenure, even fired. Their papers are rejected by major publications in their fields. The film makes a strong case for freedom of thought in scientific as in other matters.

Science and Religion. Chantilly, VA: The Teaching Company, 2006.

In video lectures, Lawrence M. Principe of Johns Hopkins University examines the relationship of science and religion from early times through the trial of Galileo, the Deist and natural theology periods, to modern fundamentalism and its response to evolution.

The Theory of Evolution: A History of Controversy. Chantilly, VA: The Teaching Company, 2002.

Edward J. Larson, whose books on the Scopes trial have been so valuable, delivers these lectures on the controversies surrounding evolution from before the time of Darwin to the present. As always, The Teaching Company programs, which frequently feature issues in religion and philosophy, are excellent.

Index

Mortier, Jeanne, 54
Moses, 2
The Mystery of Life's Origins: Reassessing Current (Thaxton), 16
Mysticism and Philosophy (Stace), 118

National Review, 6
Natural Theology (Paley), 10
Nature's Destiny (Denton), 72
Nazi party, 40, 43, 73, 110, 115
Neo-Darwinian synthesis, 12, 16, 33–35
New Geology (Price), 12, 60
Newton, Isaac, 1, 3, 8–9, 22, 25, 119, 120
Nord, Warren A., 112
Norris, J. Frank, 90

Of Pandas and People: The Central Question of Biological Origins (Davis and Kenyon), 16, 62, 104, 142
On the Origin of Species (Darwin), 11, 23, 30, 48
Orthogenesis, 33

Paley, William, 11, 22, 51, 70
Pangenesis, 33
Pascal, Blaise, 124–25
Pasteur, Louis, 64
Paul, Saint, 2
Peacocke, Arthur, 52
Peay, Governor Austin, 87
Perry, Charles, 29
Peyrere, Isaac de la, 30
The Phenomenon of Man (Teilhard de Chardin), 55–56
Pietism, 9
Piltdown Man, 54, 65, 78
Pius IX, Pope, 29
Plato, 4
Plimer, Ian, 67
Polanyi, Michael, 70
Polkinghorne, John C., 46
Postmodernism, 118, 121–23
Pound, Ezra, 115, 117
Presley, Elvis, 124
Presley, James, 94
Price, George McCready, 12–13, 60, 66
Principia Mathematica (Newton), 8
Principles of Geology (Lyell), 22

Proceedings of the Biological Society of Washington, 18, 69
Protestantism: Bibles, 14; evangelical, 14, 91; fundamentalism, 90; and law, 83, 84, 85; liberal, 101, 124; and Mormonism, 84; and public schools, 14, 110; reconciling to Darwinism, 53; resistance to Darwinism, 28–29, 96; and scientific community, 77
Public schools: Arkansas, 51; Bible classes, 85; Catholic distrust of, 14, 85; and courts, 86–105; and creationism, 67–69; 78–80; and parental rights, 13, 14, 60; and religion as program of secular instruction, 107–13
Punctuated equilibria, 77, 101

Qur'an, 49, 50, 111

Radosh, Daniel, 61–62
Rapture Ready! (Radosh), 61
The Reader's Digest, 100
Religion and the Modern Mind (Stace), 118
Riley, William Bell, 90
Robison, Carson, 92
Roman Catholicism: and Bible readings in public schools, 14, 84; and current events, 110, 112; and Darwinism, 28; and Discovery Institute, 63, 70; and eugenics, 42; and evolution, 17, 47, 53–57, 77, 91, 104; and immigration, 85; and the law, 84, 96, 100; parochial schools, 14–15
Ruse, Michael, 96–97, 102–3
Russell, Bertrand, 74, 120

Sabbath, 3, 84, 86
Sanger, Margaret, 42
Santorum, Senator Rick, 18
Schelling, Friedrick, 31
Schroeder, Gerald L., 48–49
Scientology, 126, 128
Scopes, John, 88–92
Scopes trial, 13, 78, 86, 88–96
Second Vatican Council, 53, 112
Separation of church and state, 59, 85, 97
SETI (Search for Extraterrestrial Intelligence), 65

About the Author

ALLENE PHY-OLSEN is professor emerita at Austin Peay State University. She has taught in North Africa and lectured in Sweden and Italy. She trained Peace Corps volunteers for Zaire, Liberia, and Kenya. With her husband, linguist and astronomer Frederick B. Olsen, she has traveled extensively on six continents. Now a freelance writer, her work appears in numerous reference books, including ABC-CLIO's military history titles. Her long interest in religious issues is reflected in her book *Religion and Popular Culture in America* and in reviews and articles in a variety of publications. She has also served as legal consultant and expert witness in court cases dealing with studies about religion in public schools. Her previous books include *Presenting Norma Klein, Mary Shelley,* and *Same Sex Marriage,* in Greenwood Press's Historical Guides to Controversial Issues in America series.